Surya Script Number System

Unlock the Power of Numbers:
A New Language of Symbols and Dots

Sujit Kumar Mishra

CLEVER FOX PUBLISHING
Chennai, India

Published by CLEVER FOX PUBLISHING 2025
Copyright © Sujit Kumar Mishra 2025

All Rights Reserved.
ISBN: 978-93-67077-40-5

This book has been published with all reasonable efforts taken to make the material error-free after the consent of the author. No part of this book shall be used, reproduced in any manner whatsoever without written permission from the author, except in the case of brief quotations embodied in critical articles and reviews.

The Author of this book is solely responsible and liable for its content including but not limited to the views, representations, descriptions, statements, information, opinions and references ["Content"]. The Content of this book shall not constitute or be construed or deemed to reflect the opinion or expression of the Publisher or Editor. Neither the Publisher nor Editor endorse or approve the Content of this book or guarantee the reliability, accuracy or completeness of the Content published herein and do not make any representations or warranties of any kind, express or implied, including but not limited to the implied warranties of merchantability, fitness for a particular purpose. The Publisher and Editor shall not be liable whatsoever for any errors, omissions, whether such errors or omissions result from negligence, accident, or any other cause or claims for loss or damages of any kind, including without limitation, indirect or consequential loss or damage arising out of use, inability to use, or about the reliability, accuracy or sufficiency of the information contained in this book.

Dedication

May the Surya Script Number System serve as a key to unlock new realms of understanding—where every symbol tells a story, every number holds a secret, and every equation resonates with the rhythm of the universe.

This book is dedicated to all who dare to explore beyond the ordinary,
And to the timeless wisdom that guides our search for knowledge.

INDEX

SECTION ONE

SL Number	Index	Page Number
1	Chapter # 1 - The Surya Script Part One	1-62
2	Chapter # 2 -The Surya Script Part Two Chapter	63-104
3	Chapter # 3 - The Pronunciation and Meaning of the Languages	105-108
4	Chapter # 4 – Summary in Hindi	109-116
5	Chapter #5 – The Uses of the Surya Script Number System	117-122

SECTION TWO		
1	The One-Day Secrate of Kailash	**125-126**
2	**Acknowledgment**	**127**
3	**About Author**	**128**
4	**Your Note**	**129**

INTRODUCTION PAGE

The Surya Script Number System. I would like to share some precious information for readers. the Surya Script Number System is one of simple symbolic language for general purpose used. you can also use it on the clock, calendar, data analysis, engineering works, books & notes, and calculation of numbers, etc.

This language has its pronunciation and own understanding of the languages. the reader can learn a new symbolic language without using any number system. This is more easy to understand and learn. this language uses only zero and one symbol for the whole counting.

If you reading this book you must have a basic understanding of the number system. if you know the number system you can easily understand and learn the Surya Script Number System.

In this book, we are targeting mainly English and Hindi Language for Surya Script Number System. But you can also use another foreign language of the Surya Script Number System based on the English language. like French, German, Arabic Spanish, etc. Similarly, for the Hindi language, you can also use another Indian language like Marathi, Bengali, Tamil, Telugu, Gujarati, Kannada, Ho, Udiya, etc.

SECTION ONE

CHAPTER # 1

THE SURYA SCRIPT AND LIFE PAGE

*T*he Surya Script page one is Dedicated to page one of the Surya Script Chapter one of the first symbolic languages for Surya.

The Surya Script Part One Chapter

Page Number	Symbol	Pronunciation	Meaning in English
1		OM Sa Ya One	One (1)
1		OM Sa Ya Two	Two (2)
1		OM Sa Ya Threeya	Three (3)
1		OM Sa Ya Four	Four (4)

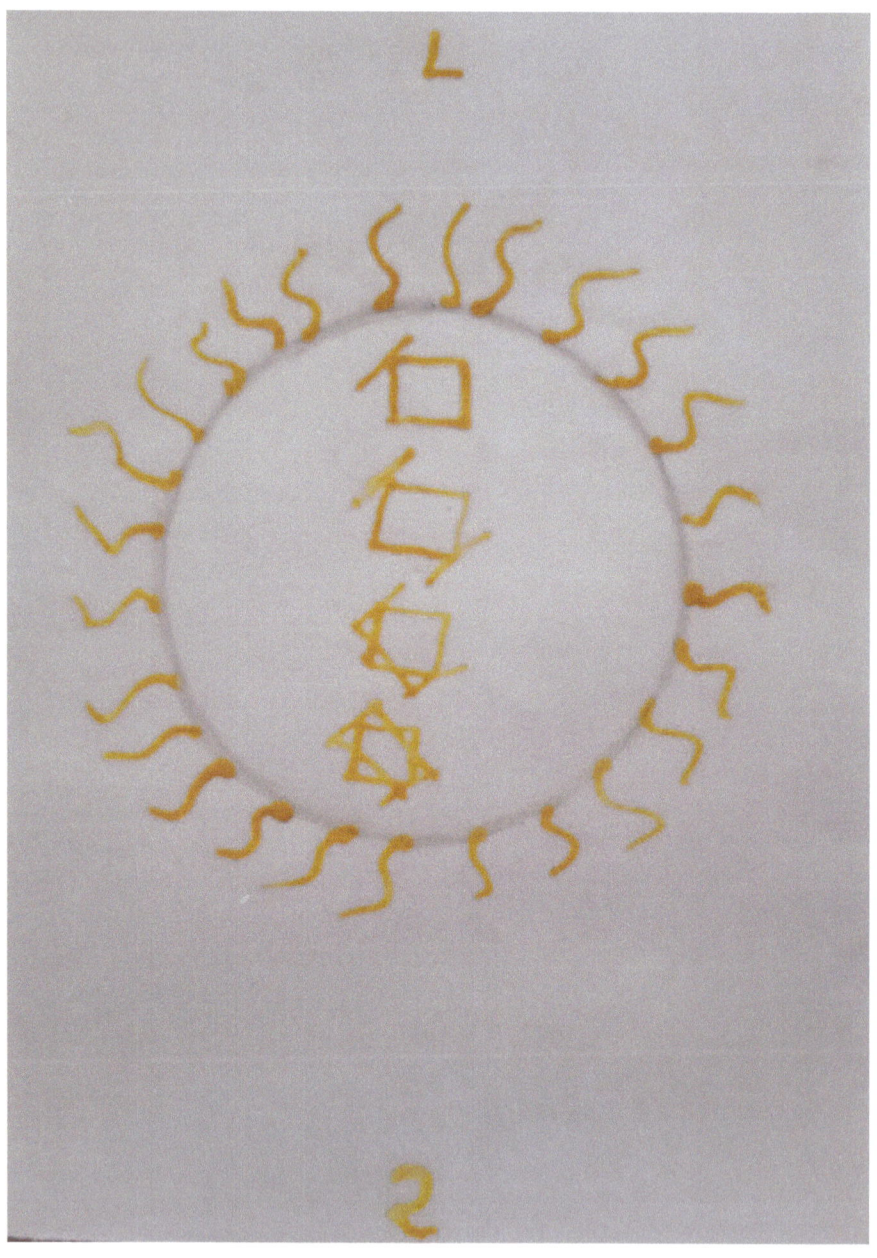

Page Number	Symbol	Pronunciation	Meaning in English
2		OM Sa Ya Five	Five (5)
2		OM Sa Ya Six	Six (6)
2		OM Sa Ya Seven	Seven (7)
2		OM Sa Ya Eight	Eight (8)

Page Number	Symbol	Pronunciation	Meaning in English
3		OM Sa Ya Nine	Nine (9)
3		OM Sa Ya Dasama	Ten (10)

Surya Script Number System

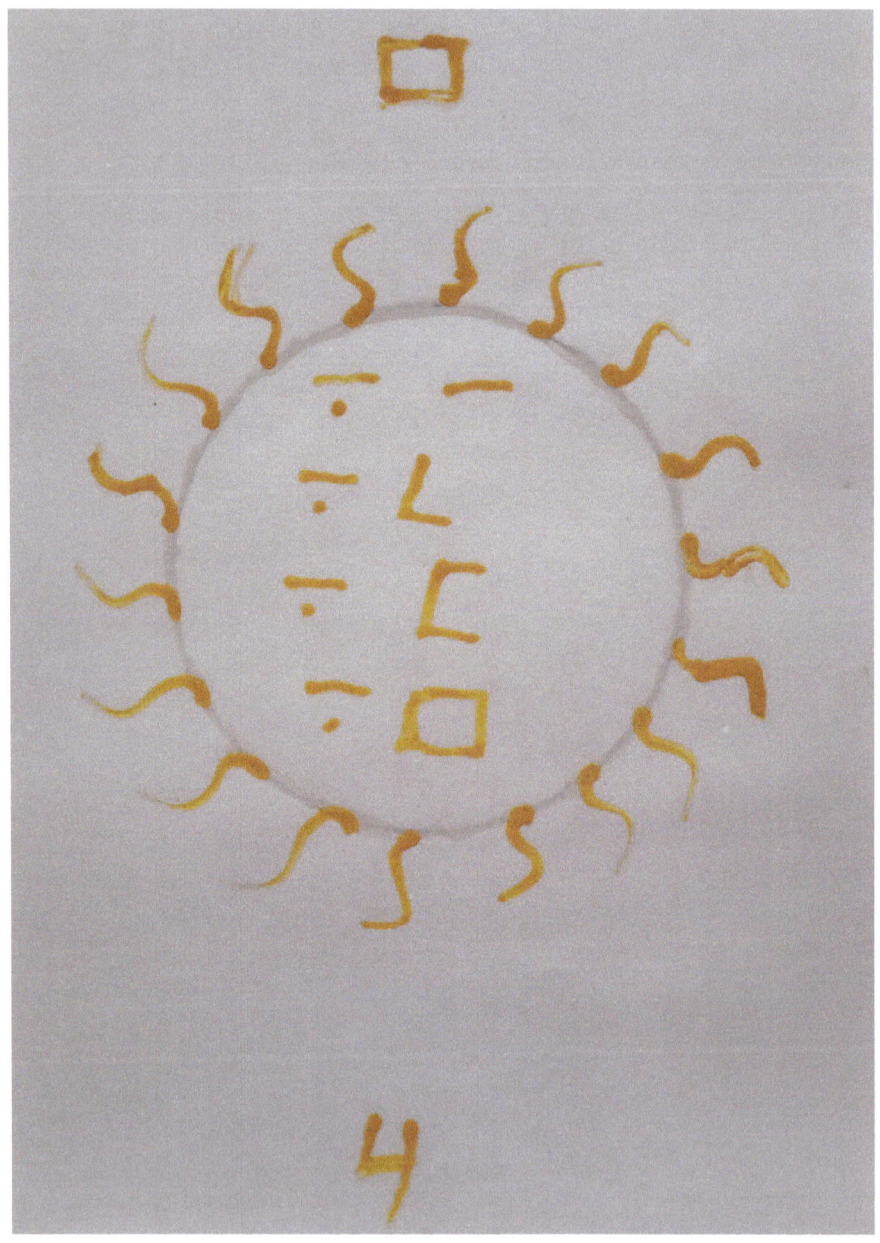

Page Number	Symbol	Pronunciation	Meaning in English
4		OM Sa Ya Eleven	Eleven (11)
4		OM Sa Ya Twelve	Twelve (12)
4		OM Sa Ya Thirteen	Thirteen (13)
4		OM Sa Ya fourteen	Fourteen(14)

Surya Script Number System

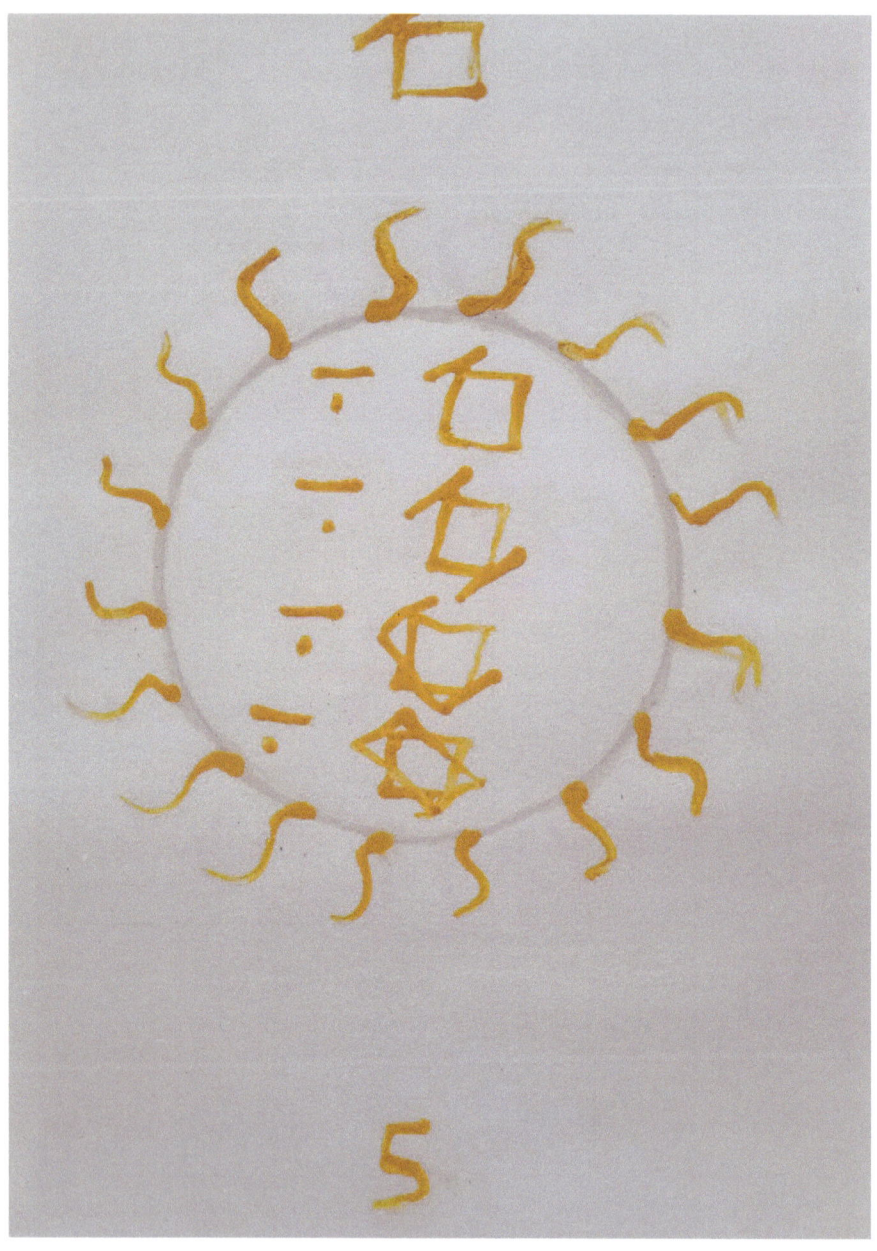

Page Number	Symbol	Pronunciation	Meaning in English
5		OM Sa Ya fifteen	Fifteen (15)
5		OM Sa Ya Sixteen	Sixteen (16)
5		OM Sa Ya Seventeen	Seventeen (17)
5		OM Sa Ya Eighteen	Eighteen (18)

Surya Script Number System

Page Number	Symbol	Pronunciation	Meaning in English
6		OM Sa Ya Ninteen	Ninteen (19)
6		OM Sa Ya Visatee	Twenty (20)

Surya Script Number System

14

Page Number	Symbol	Pronunciation	Meaning in English
7		OM Sa Ya Visatee-one	Twenty-one (21)
7		OM Sa Ya Visatee-two	Twenty-two (22)
7		OM Sa Ya Visatee -Three	Twenty-three (23)
7		OM Sa Ya Visatee-four	Twenty-four (24)

Surya Script Number System

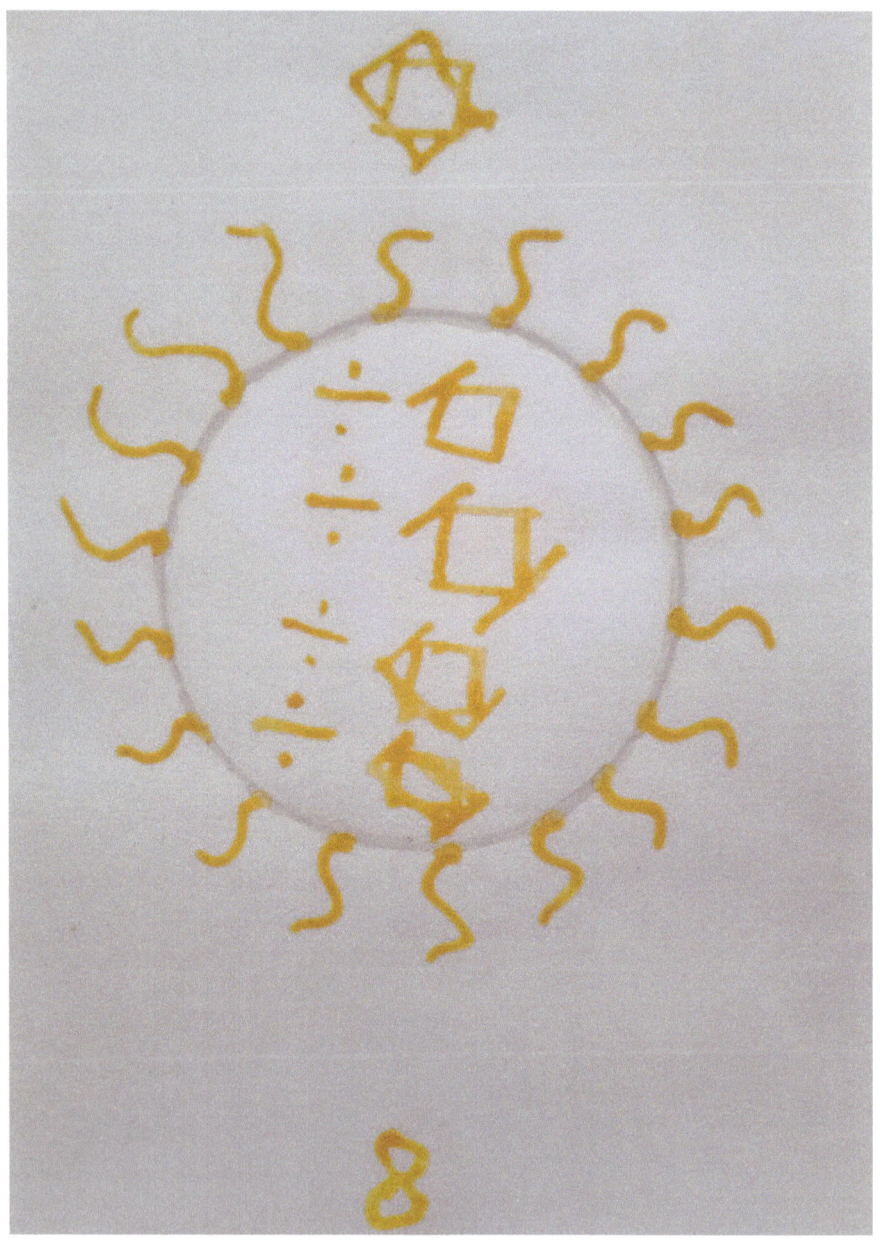

Page Number	Symbol	Pronunciation	Meaning in English
8		OM Sa Ya Visatee-five	Twenty-five (25)
8		OM Sa Ya Visatee-Six	Twenty-six (26)
8		OM Ya Visatee-Seven	Twenty-seven (27)
8		OM Sa Ya Visatee-eight	Twenty-eight (28)

Surya Script Number System

Page Number	Symbol	Pronunciation	Meaning in English
9		OM Sa Ya Visatee Nine	Twenty nine (29)
9		OM Sa Ya Trisatee	Thirty (30)

Surya Script Number System

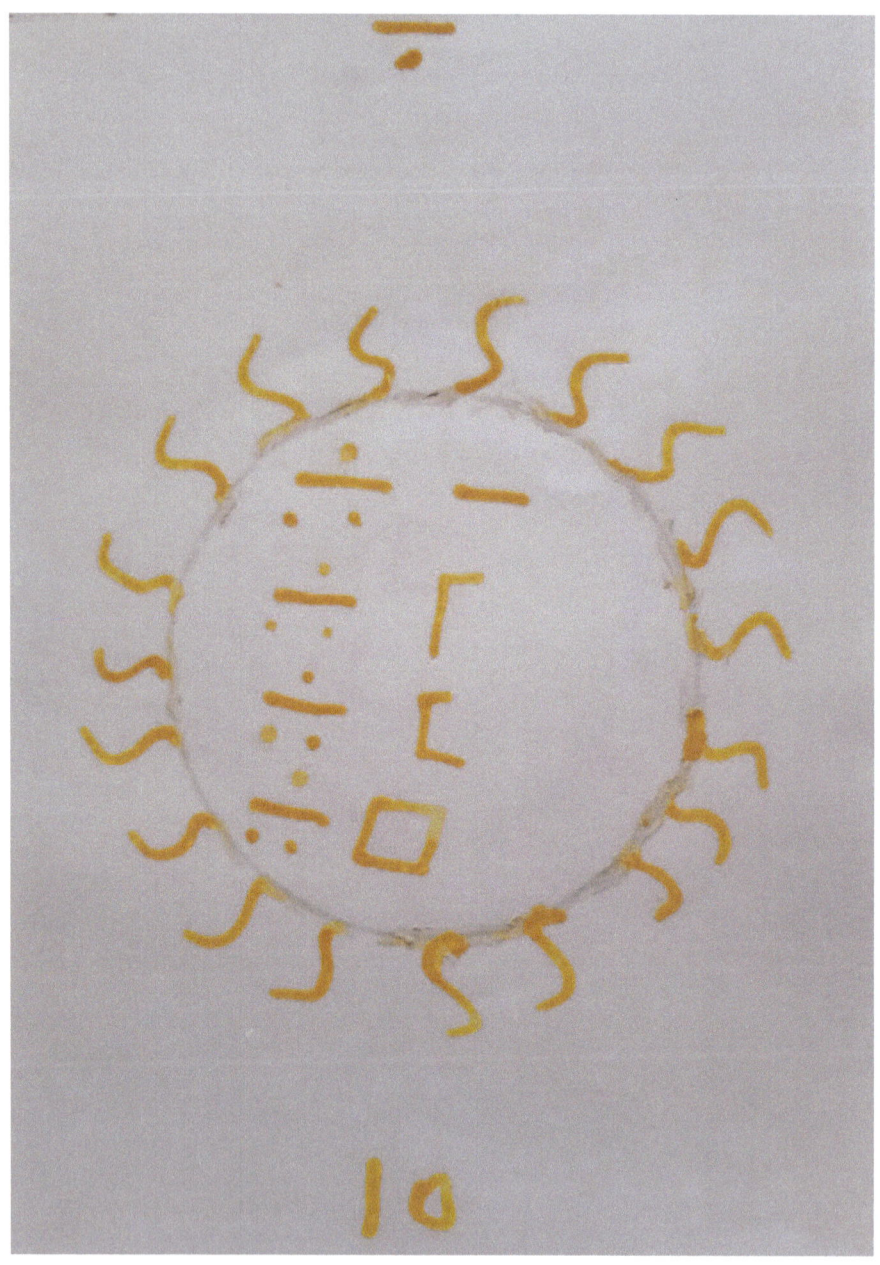

Page Number	Symbol	Pronunciation	Meaning in English
10		OM Sa Ya Trisatee-one	Thirty-one (31)
10		OM Sa Ya Trisatee -two	Thirty -two (32)
10		OM Sa Ya Trisatee -three	Thirty -three (33)
10		OM Sa Ya Trisatee -four	Thirty -four (34)

Surya Script Number System

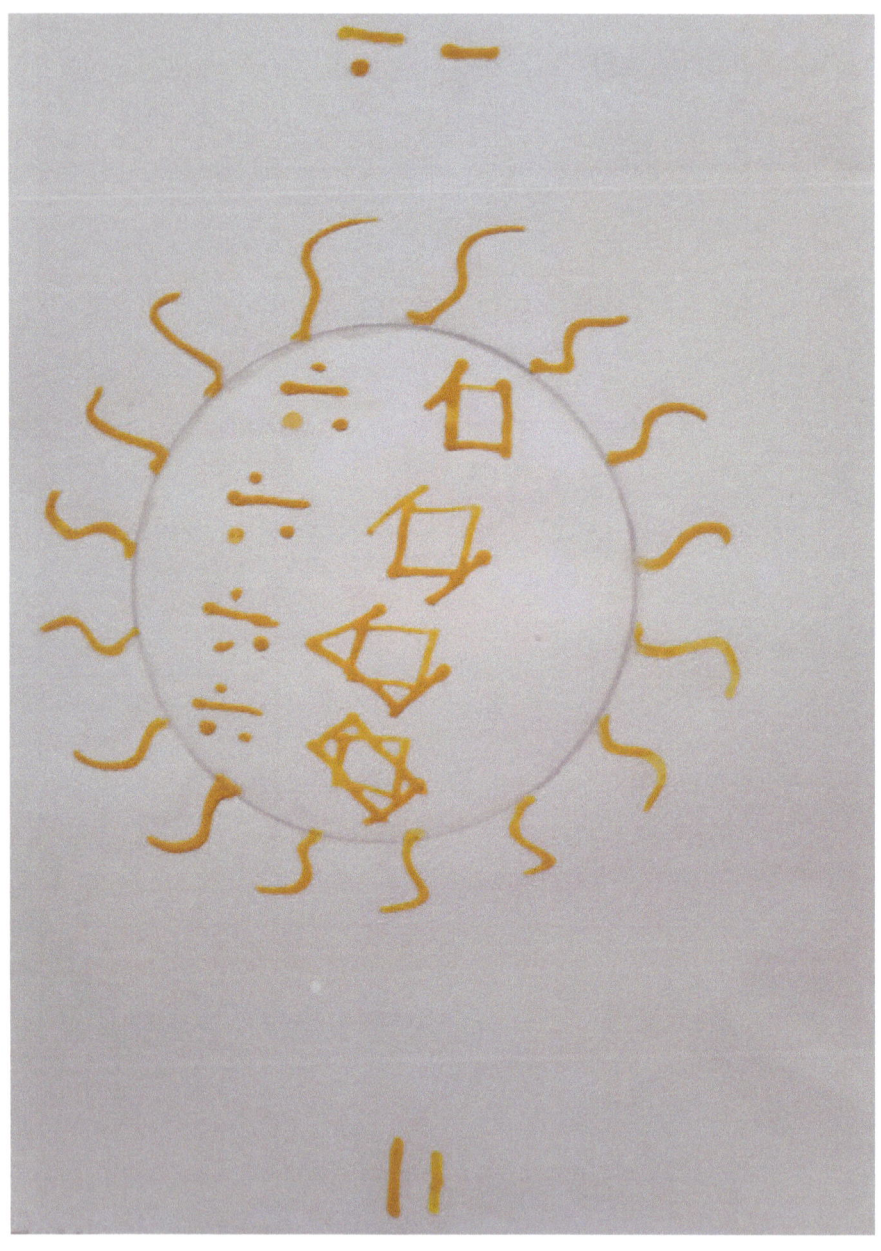

Page Number	Symbol	Pronunciation	Meaning in English
11		OM Sa Trisatee -five	Thirty -five (35)
11		OM Sa Ya Trisatee -six	Thirty -six (36)
11		MO Sa Ya Trisatee -seven	Thirty -seven (37)
11		Once Ya Trisatee -eight	Thirty -eight (38)

Surya Script Number System

Page Number	Symbol	Pronunciation	Meaning in English
12		OM Sa Ya Trisatee -Nine	Thirty -nine (39)
12		OM Sa Ya Chowaris	Forty (40)

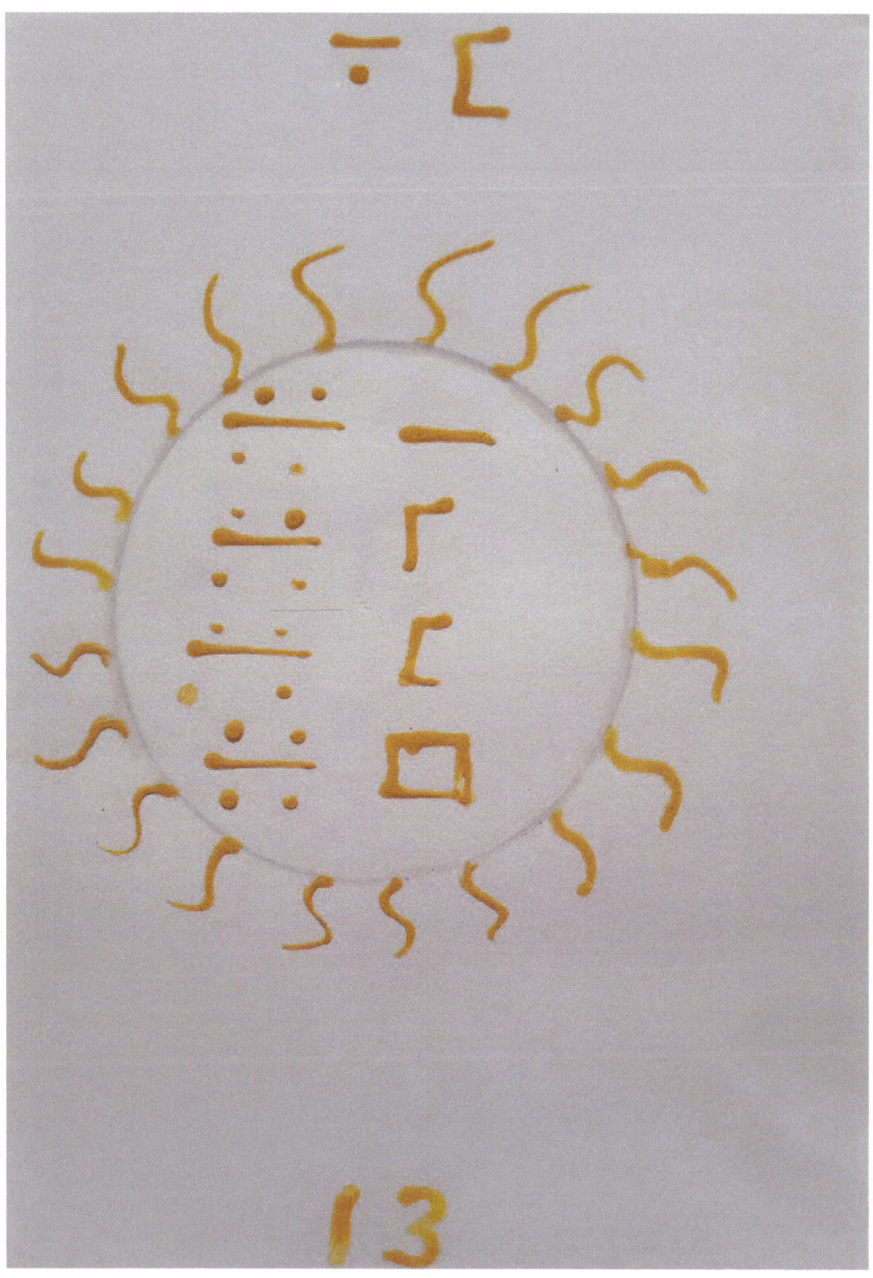

Page Number	Symbol	Pronunciation	Meaning in English
13		OM Sa Ya Chowaris -one	Forty -one (41)
13		OM Sa Ya Chowaris -two	Forty -two (42)
13		OM Sa Ya Chowaris -Three	Forty -three (43)
13		OM Sa Ya Chowaris -four	Forty-four (44)

Surya Script Number System

Page Number	Symbol	Pronunciation	Meaning in English
14		OM Sa Ya Chowaris -five	Forty -five (45)
14		OM Sa Ya Chowaris -six	Forty -six (46)
14		OM Sa Ya Chowaris -seven	Forty -seven (47)
14		OM Sa Ya Chowaris -eight	Forty -eight (48)

Surya Script Number System

Page Number	Symbol	Pronunciation	Meaning in English
15		OM Sa Ya Chowaris-Nine	Forty-nine (49)
15		OM Sa Ya Panchasat	Fifty (50)

Surya Script Number System

16

Page Number	Symbol	Pronunciation	Meaning in English
16		OM Sa Ya Punch -One	Fifty -one (51)
16		OM Sa Ya Punch -Two	Fifty -two (52)
16		OM Sa Ya Punch -three	Fifty -three (53)
16		OM Sa Ya Punch -Four	Fifty-four (54)

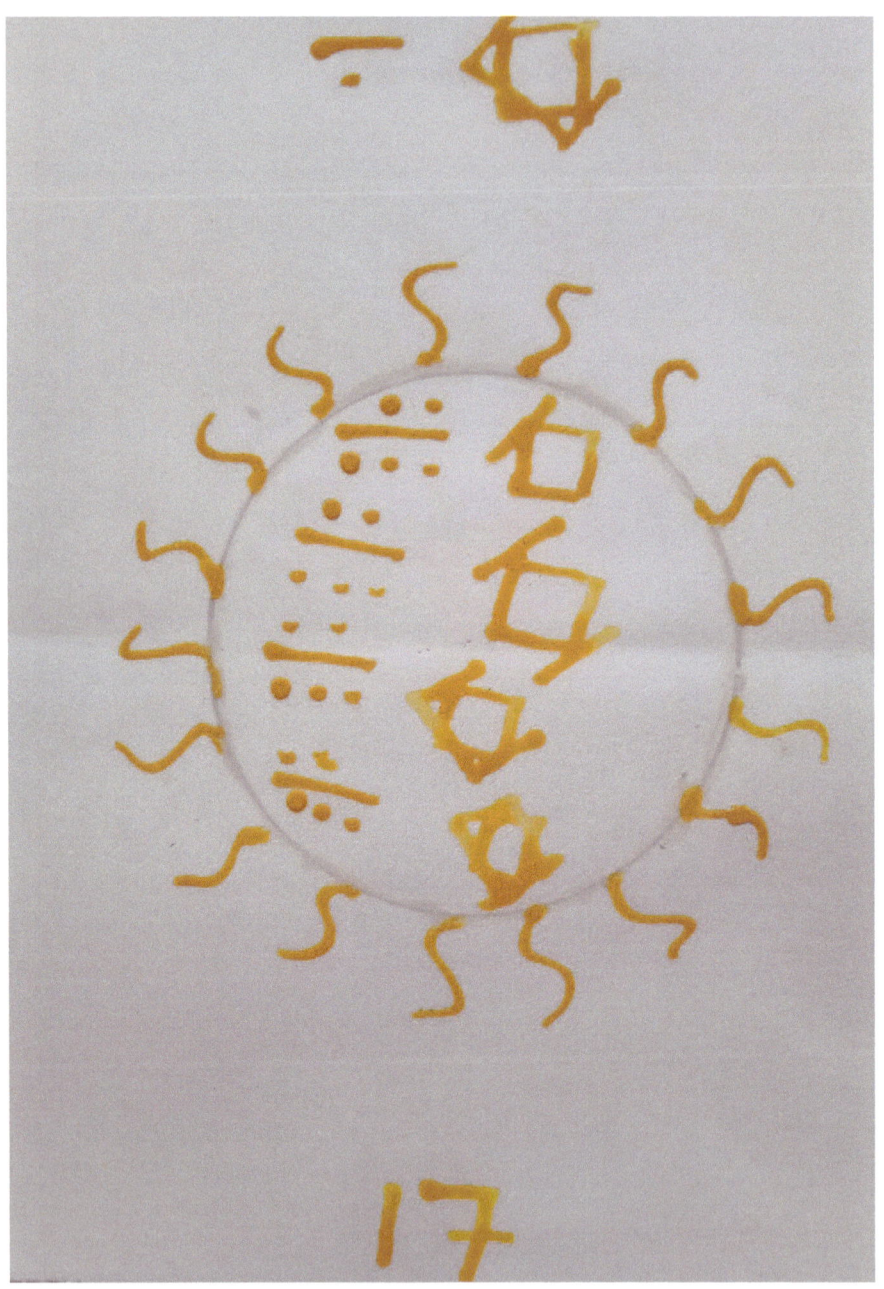

Page Number	Symbol	Pronunciation	Meaning in English
17		OM Sa Ya Punch -Five	Fifty -five (55)
17		OM Sa Ya Punch -Six	Fifty -six (56)
17		OM Sa Ya Punch -Seven	Fifty-seven (57)
17		Once Ya Punch -Eight	Fifty -eight (58)

Surya Script Number System

Page Number	Symbol	Pronunciation	Meaning in English
18		OM Sa Ya Punch-Nine	Fifty-nine (59)
18		OM Sa Ya Satth	Sixty (60)

Surya Script Number System

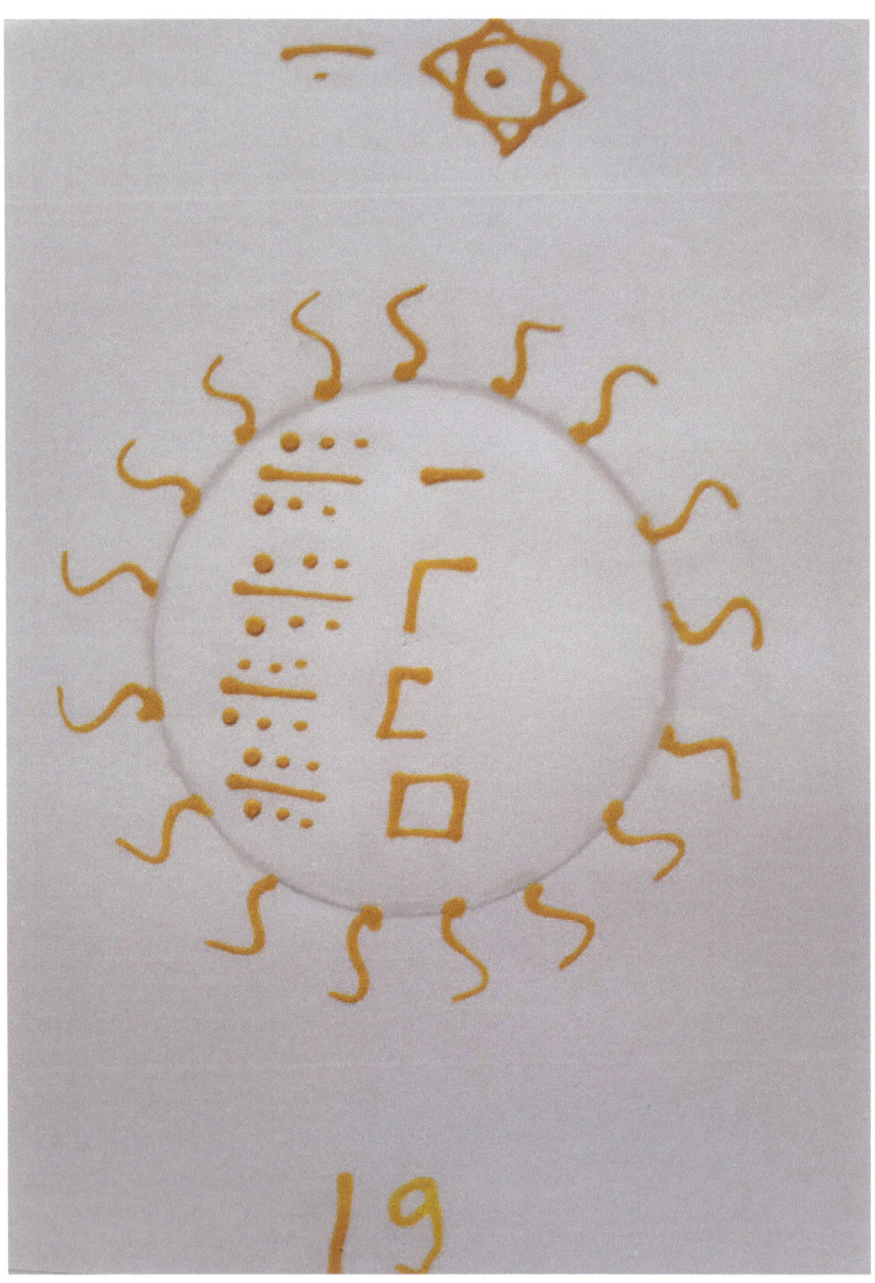

Page Number	Symbol	Pronunciation	Meaning in English
19		OM Sa Ya Satth -One	Sixty -one (61)
19		OM Sa Ya Satth -Two	Sixty -two (62)
19		OM Sa Ya Satth -Three	Sixty -three (63)
19		OM Sa Ya Satth -Four	Sixty -four (64)

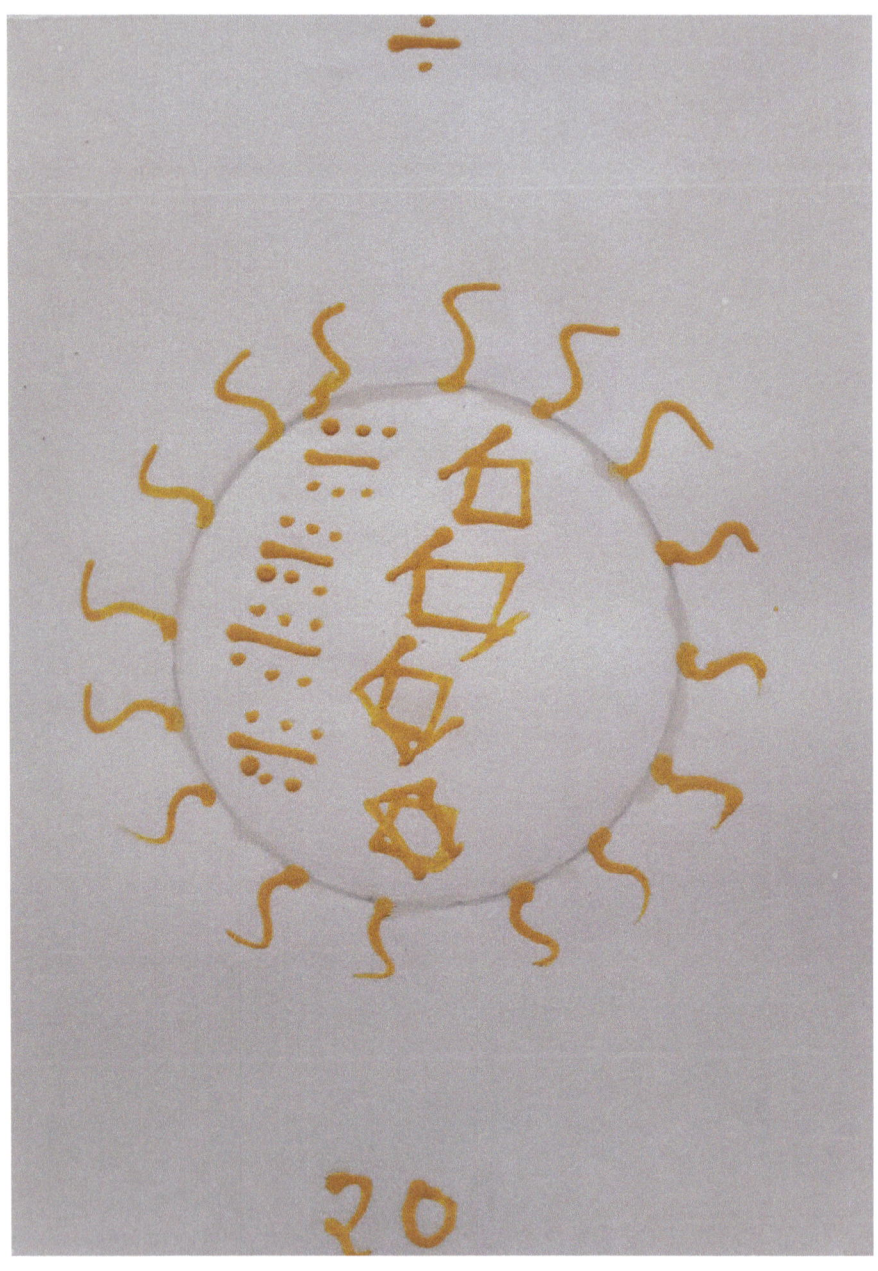

The Surya Script and Life Page

Page Number	Symbol	Pronunciation	Meaning in English
20		OM Sa Ya Satth -Five	Sixty -five (65)
20		OM Sa Ya Satth -Six	Sixty -six (66)
20		OM Sa Ya Satth -Seven	Sixty -seven (67)
20		OM Sa Ya Satth -Eight	Sixty -eight (68)

Surya Script Number System

Page Number	Symbol	Pronunciation	Meaning in English
21		OM Sa Ya Satth -Nine	Sixty-nine (69)
21		OM Sa Ya Sapti	Seventy (70)

Surya Script Number System

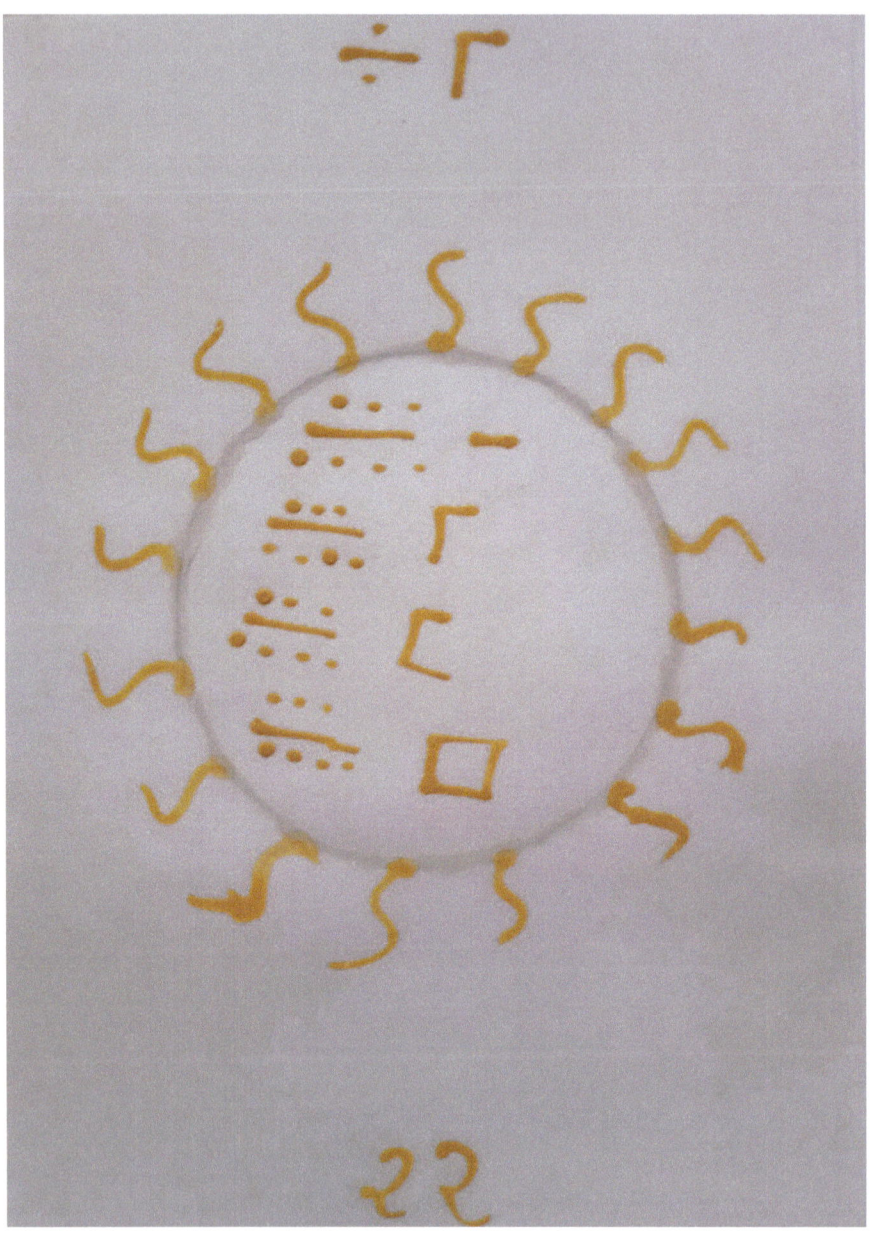

Page Number	Symbol	Pronunciation	Meaning in English
22		OM Sa Ya Sapti -One	Seventy-one (71)
22		OM Sa Ya Sapti -Two	Seventy–two (72)
22		OM Sa Ya Sapti -Three	Seventy-three (73)
22		OM Sa Ya Sapti -Four	Seventy – four (74)

Surya Script Number System

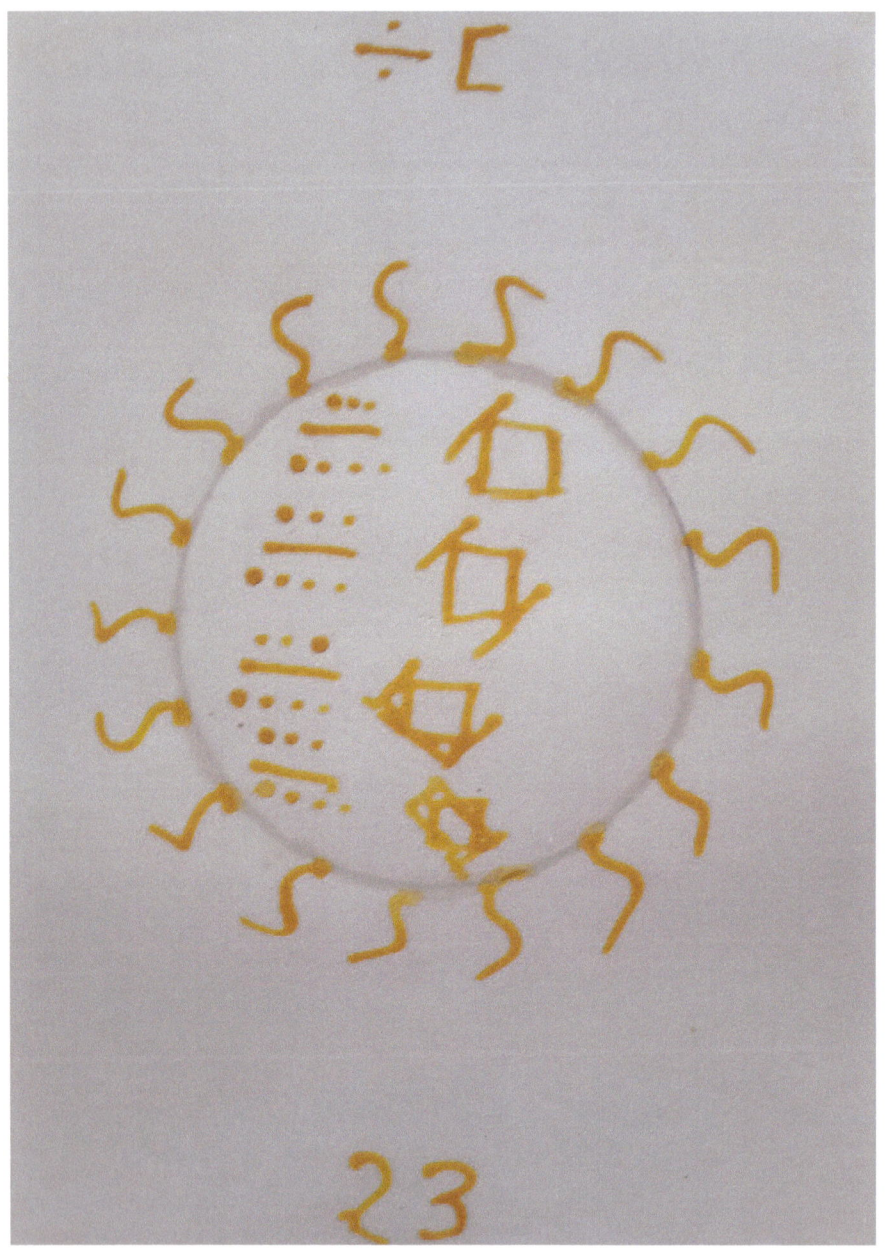

23

Page Number	Symbol	Pronunciation	Meaning in English
23		OM Sa Ya Sapti -Five	Seventy -five (75)
23		OM Sa Ya Sapti -Six	Seventy – six (76)
23		OM Sa Ya Sapti -Seven	Seventy-seven (77)
23		OM Sa Ya Sapti -Eight	Seventy – eight (78)

Surya Script Number System

24

Page Number	Symbol	Pronunciation	Meaning in English
24		OM Sa Ya Sapti Nine	Seventy-nine (79)
24		OM Sa Ya Aasit	Eighty

Surya Script Number System

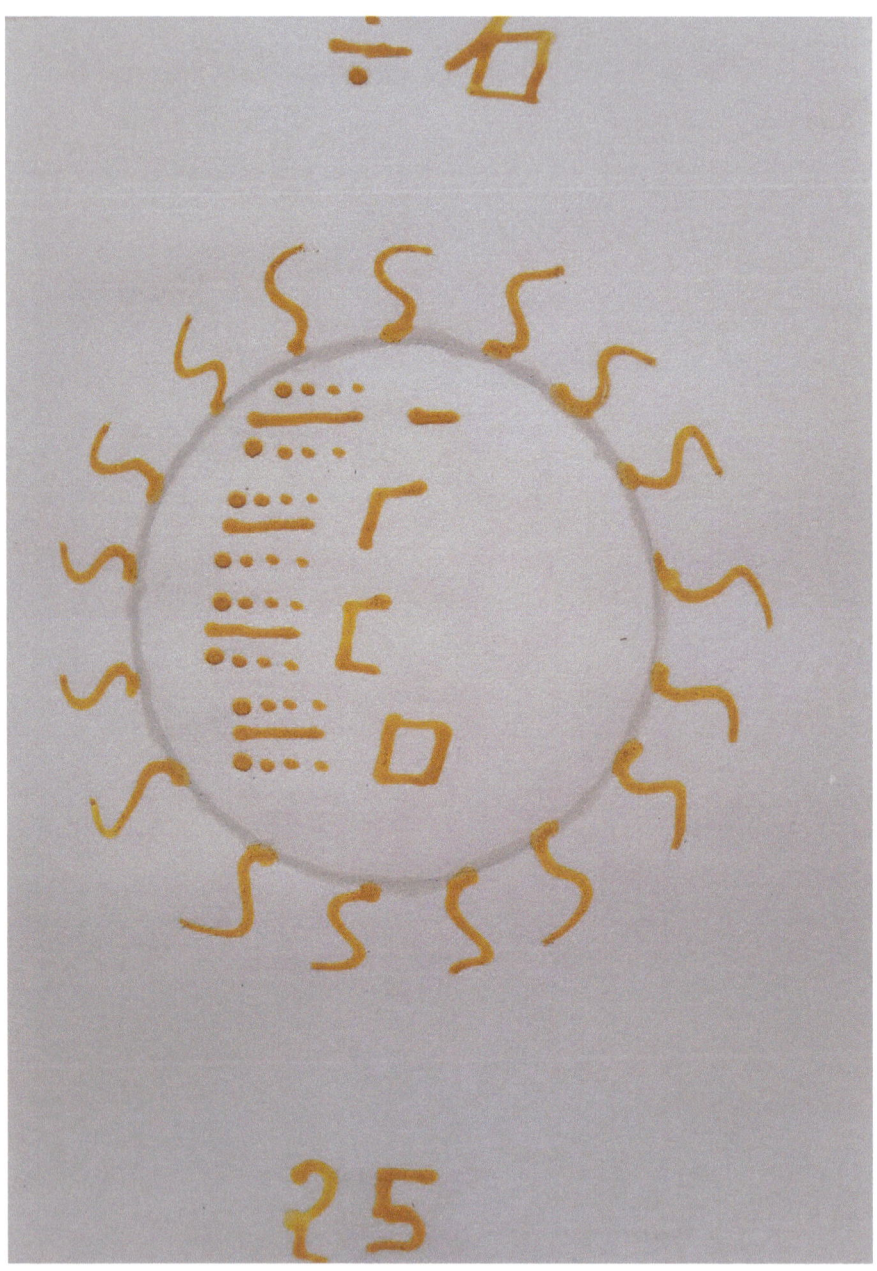

50

The Surya Script and Life Page

Page Number	Symbol	Pronunciation	Meaning in English
25		OM Sa Ya Aasit -One	Eighty-one (81)
25		OM Sa Ya Aasit -Two	Eighty – two (82)
25		OM Sa Ya Aasit -three	Eighty-three (83)
25		OM Sa Ya Aasit -Four	Eighty – four (84)

Surya Script Number System

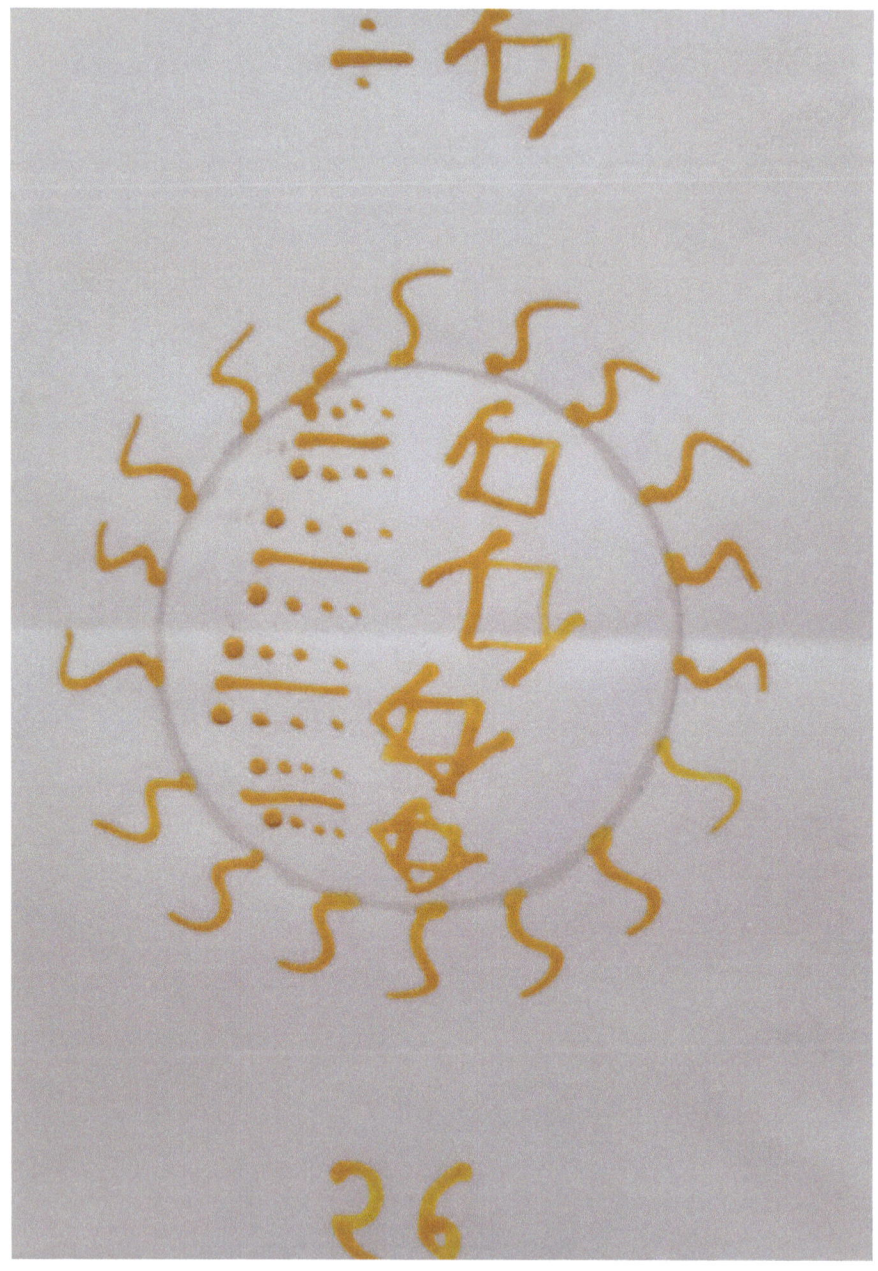

Page Number	Symbol	Pronunciation	Meaning in English
26		OM Sa Ya Aasit -Five	Eighty -five (85)
26		OM Sa Ya Aasit -Six	Eighty – six (86)
26		OM Sa Ya Aasit -Seven	Eighty -seven (87)
26		OM Sa Ya Aasit -Eight	Eighty -eight (88)

Surya Script Number System

27

Page Number	Symbol	Pronunciation	Meaning in English
27		OM Sa Ya Aasit -Nine	Eighty -Nine (89)
27		OM Sa Ya Nvati	Ninety (90)

Surya Script Number System

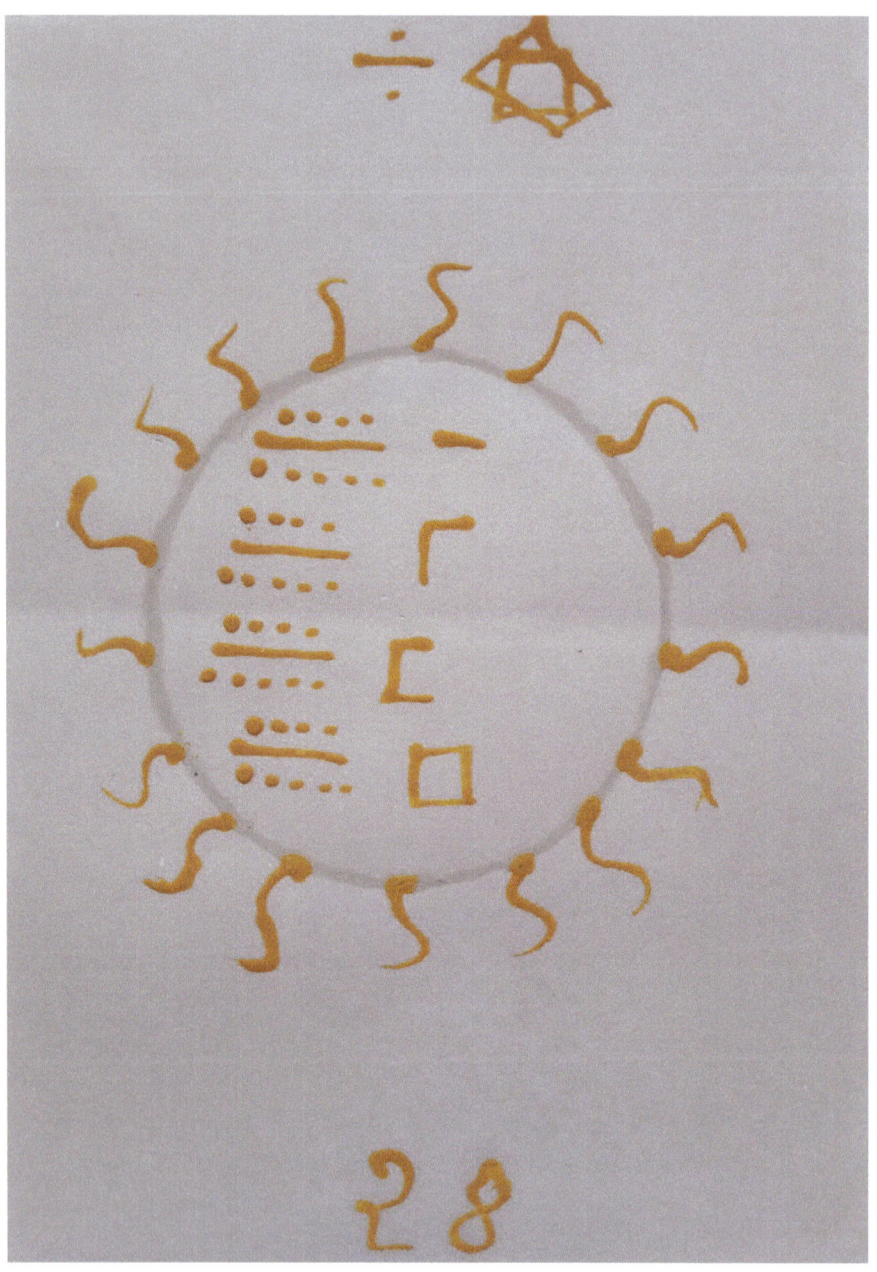

Page Number	Symbol	Pronunciation	Meaning in English
28		OM Sa Ya Nvati One	Ninety one (91)
28		OM Sa Ya Nvati Two	Ninety two (92)
28		OM Sa Ya Nvati -Three	Ninety -three (93)
28		OM Sa Ya Nvati -Four	Ninety -four (94)

Surya Script Number System

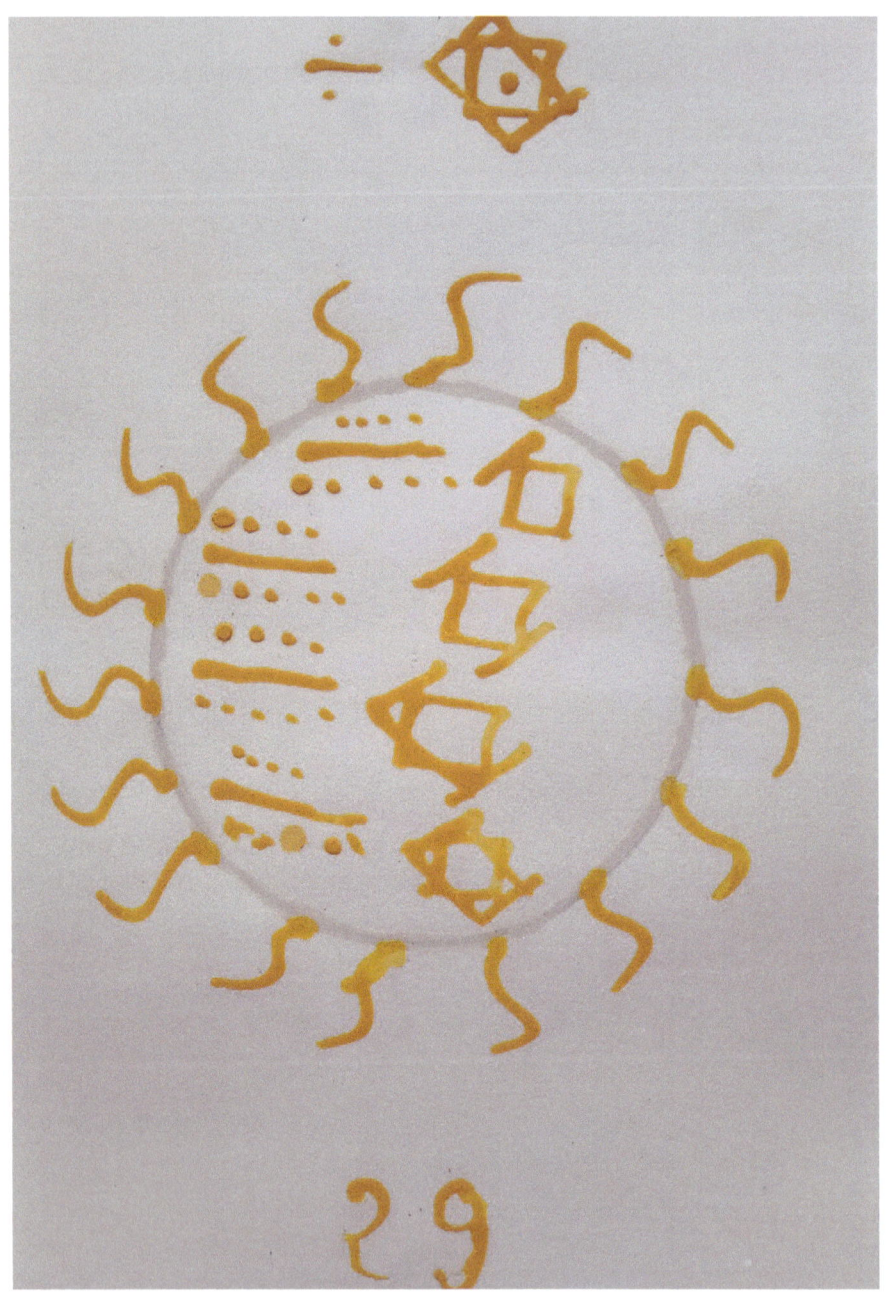

The Surya Script and Life Page

Page Number	Symbol	Pronunciation	Meaning in English
29		OM Sa Ya Nvati Five	Ninety five (95)
29		OM Sa Ya Nvati Six	Ninety six (96)
29		OM Sa Ya Nvati -Seven	Ninety -seven (97)
29		OM Sa Ya Nvati -Eight	Ninety -eight (98)

Surya Script Number System

Page Number	Symbol	Pronunciation	Meaning in English
30		OM Sa Ya Nvati -Nine	Ninety-nine (99)
30		OM Sa Ya Sattam	One hundred (100)

CHAPTER # 2

THE SURYA SCRIPT PART TWO CHAPTER

𝒯he Surya Script Symbolic Script Part One End of 100 Number with base 10.

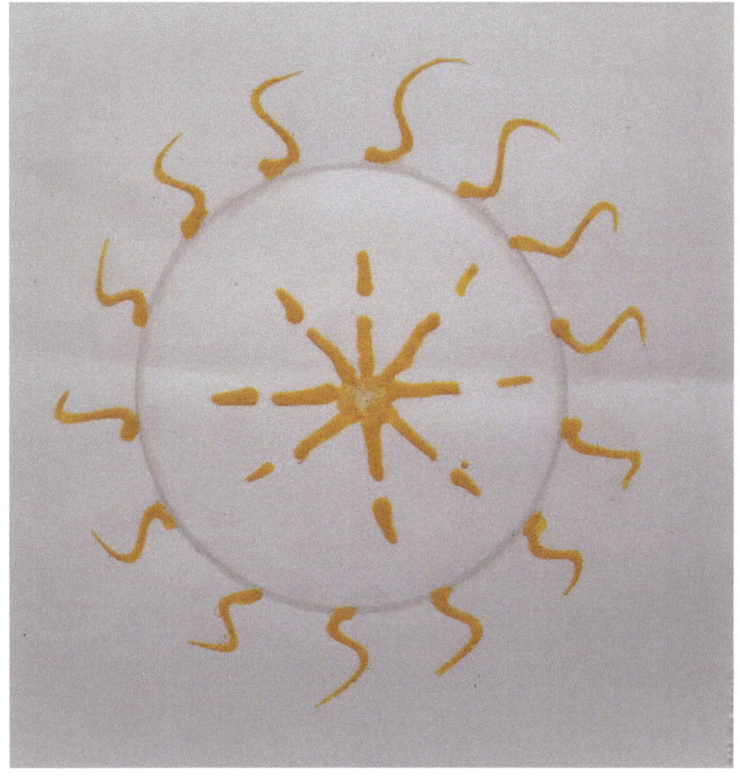

The Surya Script Number beyond 100 Start

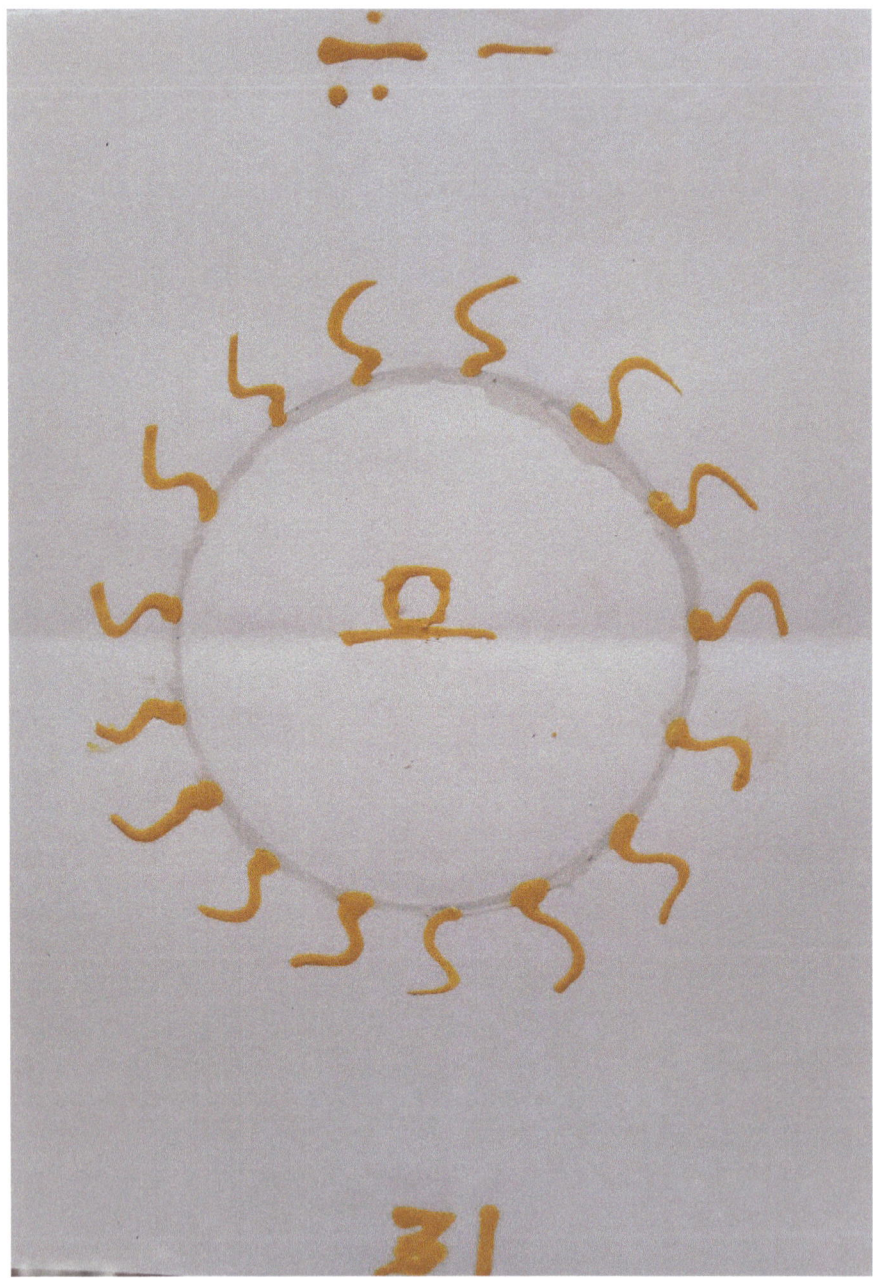

This number is used only after the 100-completion count; it's an alternative number after the 100-completion count.

Page Number	Symbol	Pronunciation	Meaning in English
31		OM Sa Ya Satam	Hundred (100)
31		OM Sa Ya Satam	10^2

Surya Script Number System

Page Number	Symbol	Pronunciation	Meaning in English
32		OM Sa Ya Satam One	One hundred one (101)
32		OM Sa Ya Satam Two	One hundred two (102)
32		OM Sa Ya Satam Three	One hundred three (103)
32		OM Sa Ya Satam Four	One hundred four (104)

Surya Script Number System

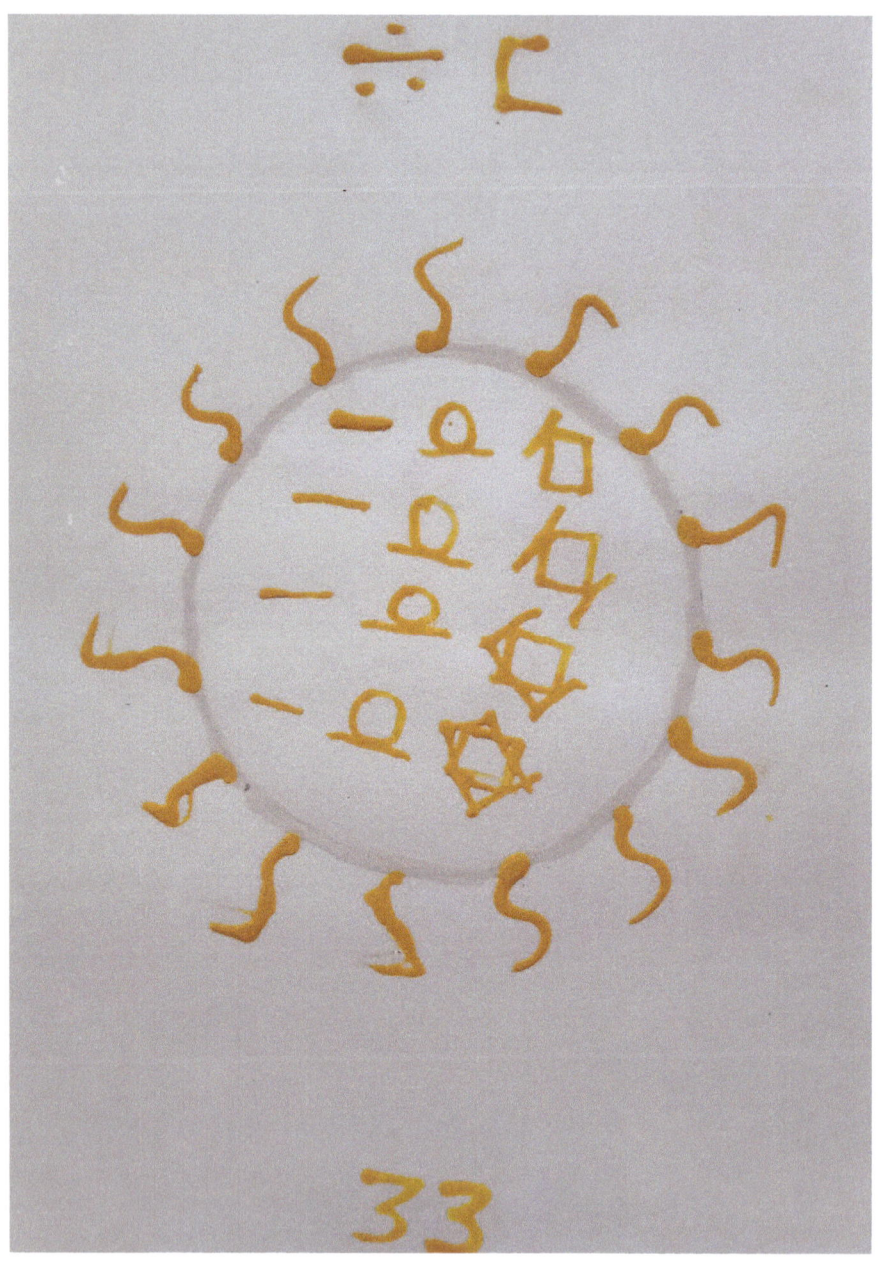

33

Page Number	Symbol	Pronunciation	Meaning in English
33		OM Sa Ya Satam Five	One hundred five (105)
33		OM Sa Ya Satam Six	One hundred six (106)
33		OM Sa Ya Satam Seven	One hundred-seven (107)
33		OM Sa Ya Satam Eight	One hundred-eight (108)

Surya Script Number System

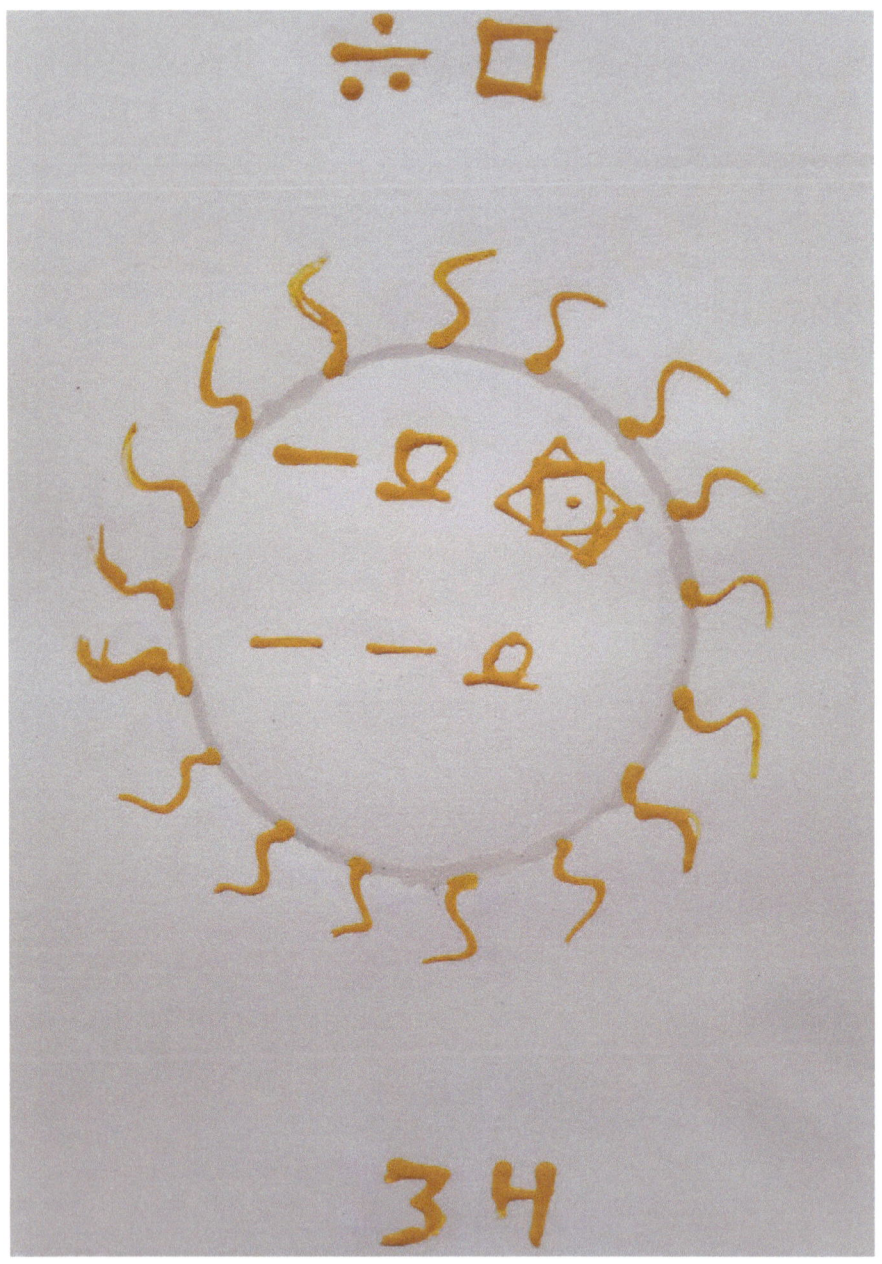

70

Page Number	Symbol	Pronunciation	Meaning in English
34		OM Sa Ya Satam Nine	One hundred-Nine (109)
34		OM Sa Ya Satam Dasam	One hundred-Ten (110)

Surya Script Number System

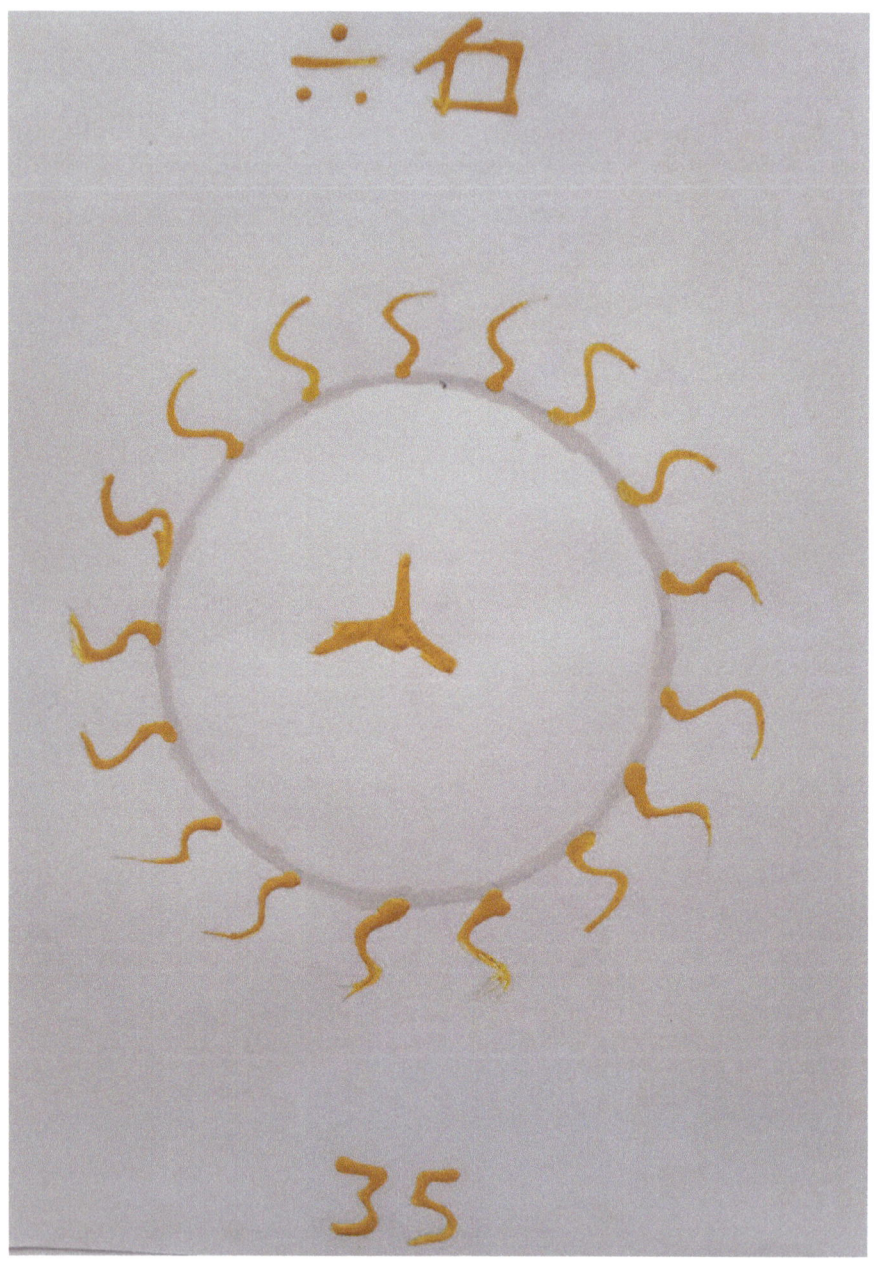

Page Number	Symbol	Pronunciation	Meaning in English
35		OM Sa Ya Hazar	One thousand
35	Number	10 X 10 X 10	1000

Surya Script Number System

Page Number	Symbol	Pronunciation	Meaning in English
36		OM Sa Ya Des Hazar	Ten Thousand
36	Number	10 X 10 X 10 X 10	10000

Surya Script Number System

Page Number	Symbol	Pronunciation	Meaning in English
37		OM Sa ya Lakh	Hundred Thousand
37	Number	10 X 10 X 10 X 10 x 10	1,00,000

Surya Script Number System

Page Number	Symbol	Pronunciation	Meaning in English
38		OM Sa ya Des lakh	One Million
38	Number	10^6	10,00,000

Surya Script Number System

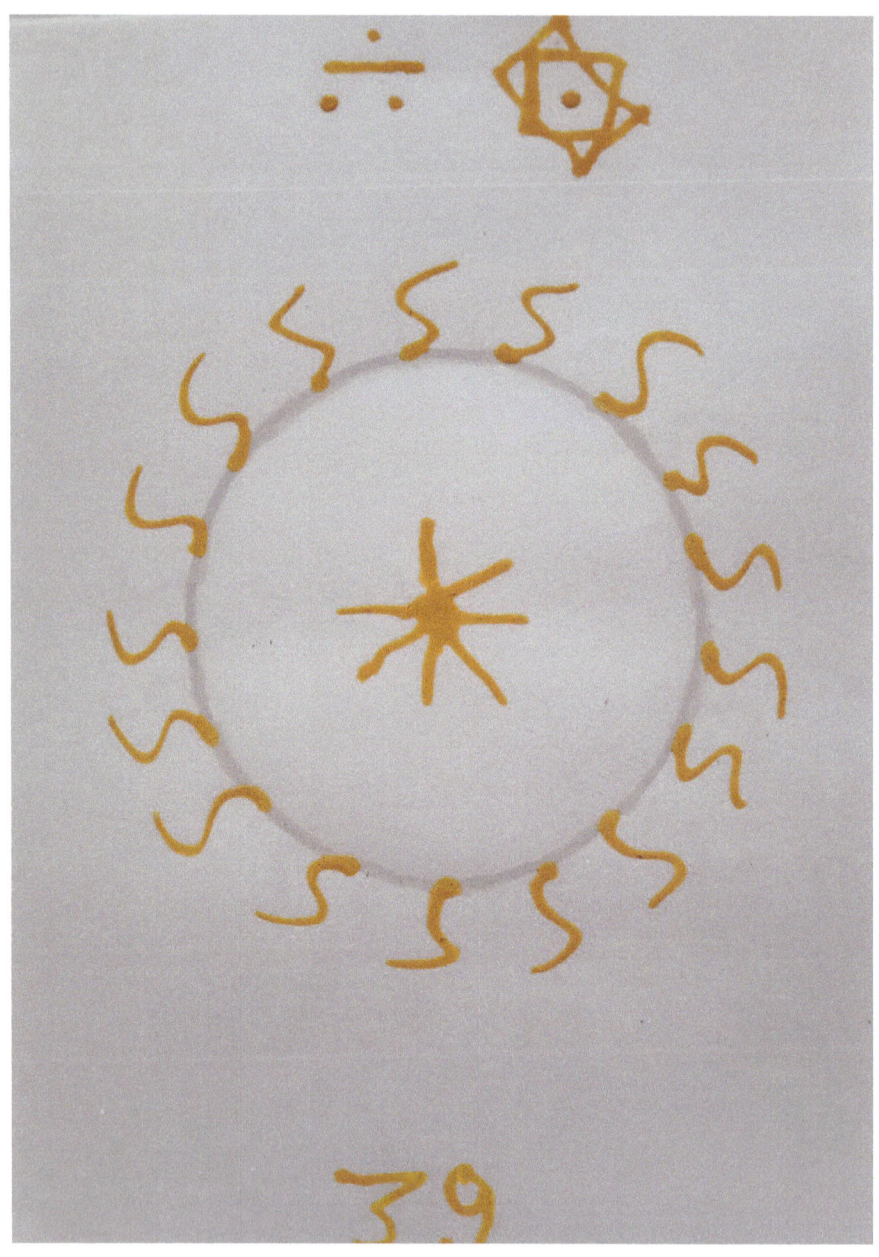

39

Page Number	Symbol	Pronunciation	Meaning in English
39		OM Sa ya Crore	Ten Million
39	Number	10^7	1,00,00,000

Surya Script Number System

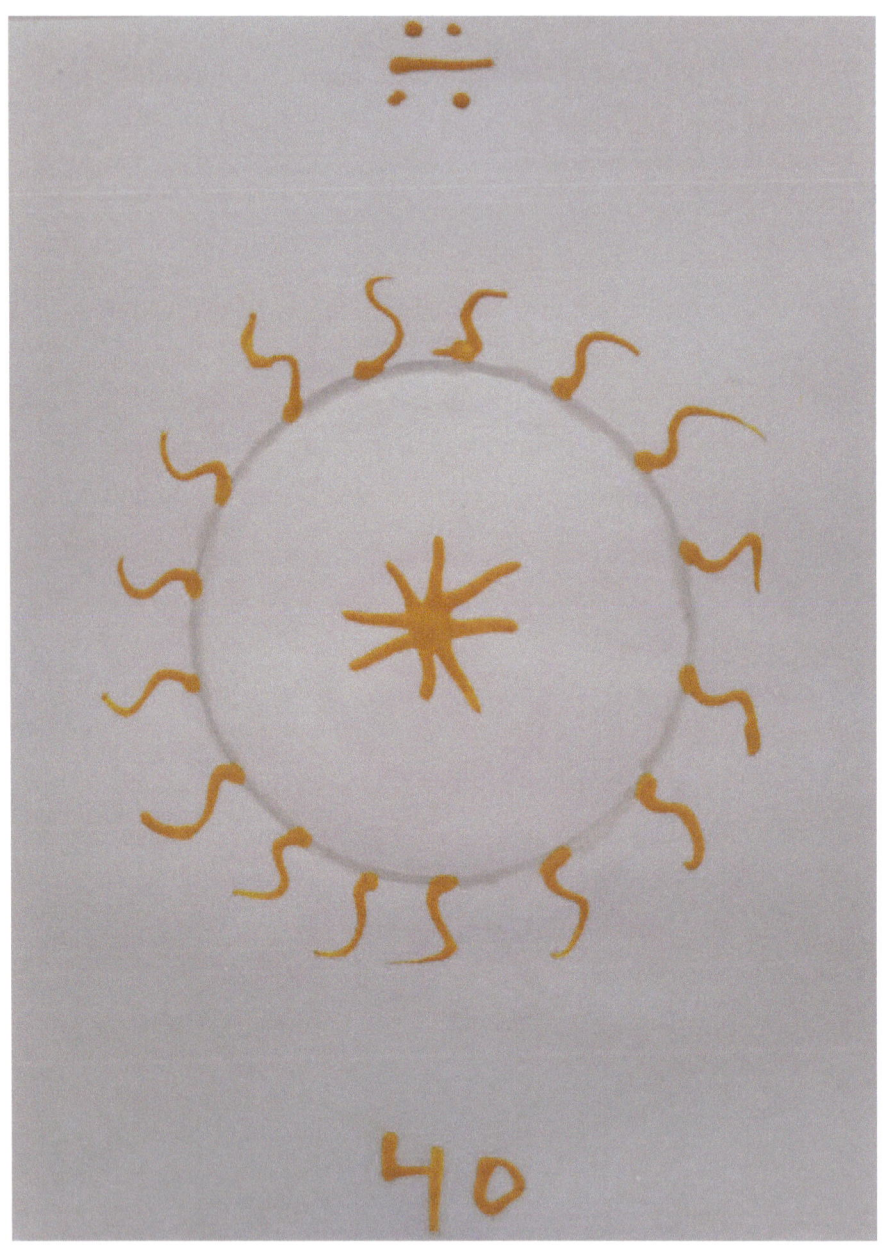

Page Number	Symbol	Pronunciation	Meaning in English
40		OM Sa yDes Crore	Hundred Million
40	Number	10^8	10,00,00,000

Surya Script Number System

Page Number	Symbol	Pronunciation	Meaning in English
41		OM Sa Ya Arab	One Billion
41	Number	10^9	1,00,00,00,000

Surya Script Number System

42

Page Number	Symbol	Pronunciation	Meaning in English
42		OM Sa ya Des Arab	Ten Billion
42	Number	10^{10}	10,00,00,00,000

Surya Script Number System

43

Page Number	Symbol	Pronunciation	Meaning in English
43		OM Sa ya Kharab	Hundred Billion
43	Number	10^{11}	1,00,00,00,00,000

Surya Script Number System

44

Page Number	Symbol	Pronunciation	Meaning in English
44		OM Sa ya ten Kharab	One Tillion
44	Number	10^{12}	10,00,00,00,00,000

Surya Script Number System

45

Page Number	Symbol	Pronunciation	Meaning in English
45		OM Sa ya Neel	Ten Trillion
45	Number	10^{13}	1,00,00,00,00,00,000

Page Number	Symbol	Pronunciation	Meaning in English
46		OM Sa ya Des Neel	Hundred Trillion
46	Number	10^{14}	10,00,00,00,00,00,000

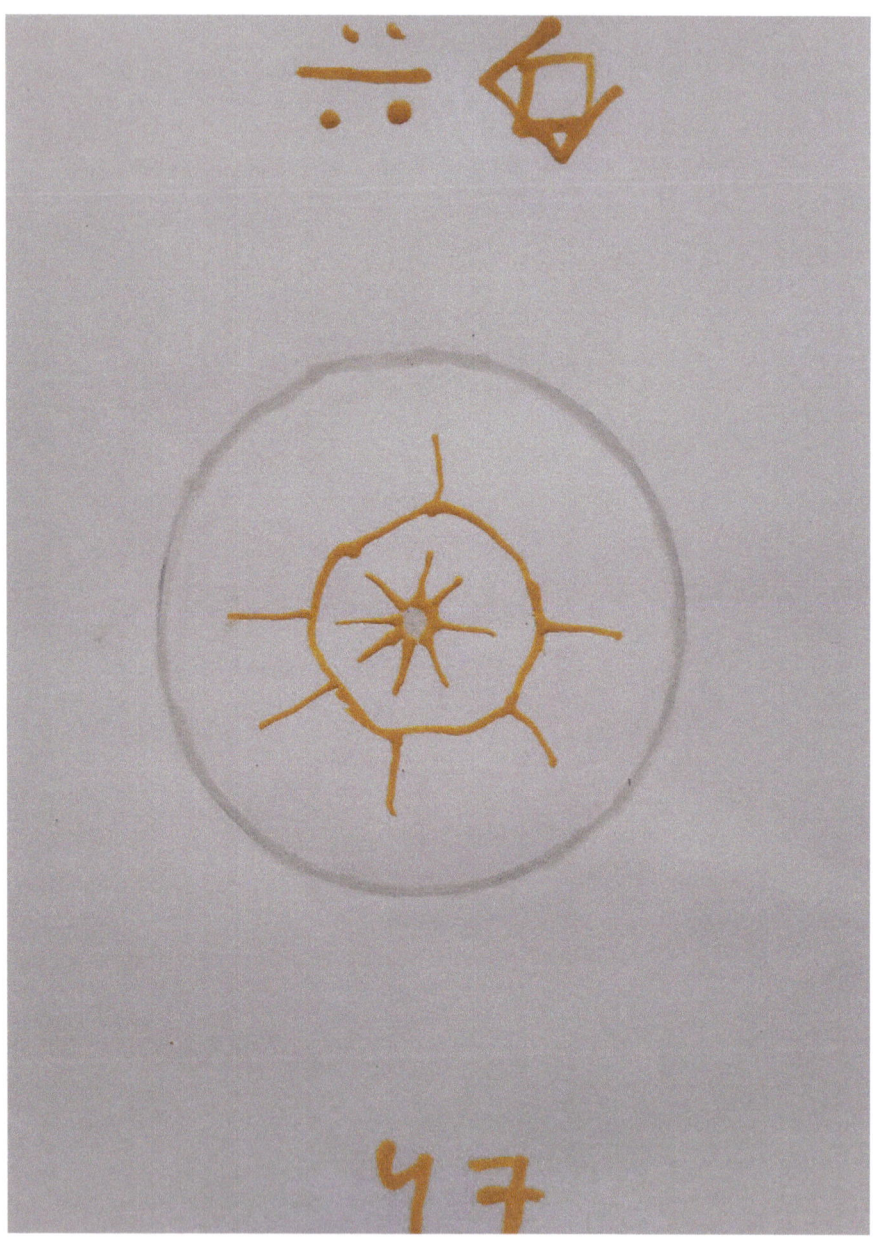

Page Number	Symbol	Pronunciation	Meaning in English
47		OM Sa ya Padam	One Quadrillion
47	Number	10 ^{15}	1,00,00,00,00,00,00,000

Surya Script Number System

Page Number	Symbol	Pronunciation	Meaning in English
48		OM ya Anta One	Ten Quadrillion
48	Number	10^{16}	10,00,00,00,00,00,00,000

Surya Script Number System

100

Page Number	Symbol	Pronunciation	Meaning in English
49		OM Sa ya Anta Do	Hundred
49	Number	10^{17}	1,00,00,00,00,00,00,00,000

Surya Script Number System

Page Number	Symbol	Pronunciation	Meaning in English
50		Once ya Anta Triy	Quadrillion
50	Number	10 ^{18}	10,00,00,00,00,00,00,00,000

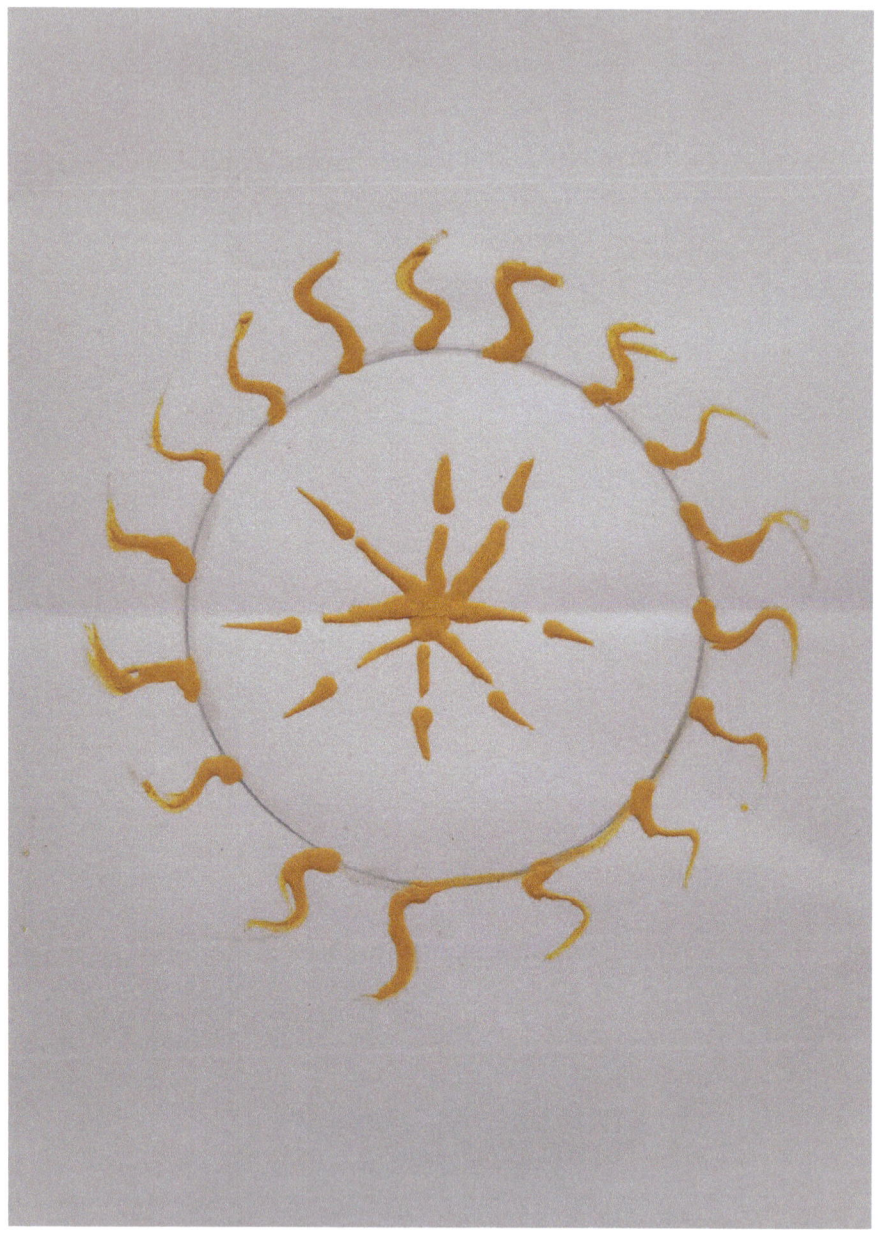

The Surya Script ends for our count number

CHAPTER # 3

THE PRONUNCIATION AND MEANING OF LANGUAGES.

The Surya Script with Three Sun Symbol with Base 10 Number system. It's very easy to understand the language of the **Surya Script Number System.** Every first Sun Symbol indicated 4 numbers with one Sun Circle Symbol. Next, the second Symbol was 4 numbers with one Sun Circle Symbol, and the last one was 2 numbers with one Sun Circle Symbol. So the Total number is 10. When going to the higher number system the Sun Symbol system is not fixed it will be changed as per Script languages. It changes with time and the necessity of work.

Here we would like to show how to pronounce the Surya Script and write in your language. This book is dedicated to the first Indian language pronunciation of Surya Script. The Language is stated below as a summary of different languages—for example Hindi.

You can also use different Indian languages based on Hindi like Marathi, Bengali, Gujarati, Punjabi, Urdu, Tamil, Kanad, Malayalam, Ho, etc., or Foreign Languages you can use based on English Languages of the **Surya Script Number System.**

SURYA SCRIPT SUMMARY
(HINDI)

CHAPTER # 4

SUMMARY IN HINDI

Surya Script Number System in the Hindi language is Based on 10 Numbers.

आकृति	उच्चारण	अर्थ
—	उमसाया एक	१
⌐	उमसाया दो	२
⊔	उमसाया तीन	३
☐	उमसाया चार	४
⬠	उमसाया पाँच	५
⬠	उमसाया छ	६
⬡	उमसाया सात	७
✴	उमसाया आठ	८
✴	उमसाया नौ	९
∴	उमसाया दसमा	१०

Surya Script Number System

आकृती	उच्चारण	अर्थ
÷ -	उमसांख्या ग्यारह	११
÷ L	उमसांख्या बारह	१२
÷ ⊔	उमसांख्या तेरह	१३
÷ ☐	उमसांख्या चौदह	१४
÷ ⌂	उमसांख्या पंद्रह	१५
÷ ⌂	उमसांख्या सोलह	१६
÷ ✧	उमसांख्या सत्रह	१७
÷ ✦	उमसांख्या अठारह	१८
÷ ✦	उमसांख्या उन्नीस	१९
∴	उमसांख्या बीस	२०

आकृती	उच्चारण	अर्थ
÷ -	उमशांख्या इक्कीस	२१
÷ L	उमशांख्या बाईस	२२
÷ ⊔	उमशांख्या तेईस	२३
÷ ☐	उमशांख्या चौबीस	२४
÷ ⌂	उमशांख्या पच्चीस	२५
÷ ⌂	उमशांख्या छब्बीस	२६
÷ ✧	उमशांख्या सत्ताईस	२७
÷ ✦	उमशांख्या अट्ठाईस	२८
÷ ✦	उमशांख्या उन्तीस	२९
∴	उमशांख्या तीस	३०

Summary in Hindi

आकृति	उच्चारण	अर्थ
∴ –	उमशांयां त्रिमही एक	३१
∴ L	उमशांयां त्रिमही दो	३२
∴ ⌐	उमशांयां त्रिमही तीन	३३
∴ □	उमशांयां त्रिमही चार	३४
∴ ⬠	उमशांयां त्रिमही पाँच	३५
∴ ⬡	उमशांयां त्रिमही ह:	३६
∴ ⬢	उमशांयां त्रिमही सात	३७
∴ ✡	उमशांयां त्रिमही आठ	३८
∴ ✦	उमशांयां त्रिमही नौ	३९
∴	उमशांयां चौगारिस	४०

आकृति	उच्चारण	अर्थ
∴ –	उमशांयां चौगारीस एक	४१
∴ L	उमशांयां चौगारीस दो	४२
∴ ⌐	उमशांयां चौगारीस तीन	४३
∴ □	उमशांयां चौगारीस चार	४४
∴ ⬠	उमशांयां चौगारीस पाँच	४५
∴ ⬡	उमशांयां चौगारीस ह:	४६
∴ ⬢	उमशांयां चौगारीस सात	४७
∴ ✡	उमशांयां चौगारीस आठ	४८
∴ ✦	उमशांयां चौगारीस नौ	४९
∴	उमशांयां पञ्परगत्	५०

आकृती	उच्चारण	अर्थ
⋮⋮ −	उमशांया पञ्चशत् एक	५१
⋮⋮ L	उमशांया पञ्चशत् दो	५२
⋮⋮ ⨆	उमशांया पञ्चशत् तीन	५३
⋮⋮ ▢	उमशांया पञ्चशत् चार	५४
⋮⋮ ⟁	उमशांया पञ्चशत् पाँच	५५
⋮⋮ ⬠	उमशांया पञ्चशत् दः	५६
⋮⋮ ⬡	उमशांया पञ्चशत् सात	५७
⋮⋮ ✶	उमशांया पञ्चशत् आठ	५८
⋮⋮ ✹	उमशांया पञ्चशत् नौ	५९
⋮⋮	उमशांया साठ	६०

आकृती	उच्चारण	अर्थ
⋮⋮⋮ −	उमशांया साठ एक	६१
⋮⋮⋮ L	उमशांया साठ दो	६२
⋮⋮⋮ ⨆	उमशांया साठ तीन	६३
⋮⋮⋮ ▢	उमशांया साठ चार	६४
⋮⋮⋮ ⟁	उमशांया साठ पाँच	६५
⋮⋮⋮ ⬠	उमशांया साठ दः	६६
⋮⋮⋮ ⬡	उमशांया साठ सात	६७
⋮⋮⋮ ✶	उमशांया साठ आठ	६८
⋮⋮⋮ ✹	उमशांया साठ नौ	६९
⋮⋮⋮	उमशांया सप्तती	७०

Summary in Hindi

आहुती	उच्चारण	अर्थ
⋮ –	उमशांया सप्ती एक	६१
⋮ L	उमशांया सप्ती दो	६२
⋮ Ụ	उमशांया सप्ती तीन	६३
⋮ □	उमशांया सप्ती चार	६४
⋮ ⌂	उमशांया सप्ती पाँच	६५
⋮ ⌂	उमशांया सप्ती ६ः	६६
⋮ ⬠	उमशांया सप्ती सात	६७
⋮ ✶	उमशांया सप्ती आळ	६८
⋮ ✶	उमशांया सप्ती नौ	६९
⋮⋮	उमशांया अप्ती	८०

आहुती	उच्चारण	अर्थ
⋮⋮ –	उमशांया अप्ती एक	८१
⋮⋮ L	उमशांया अप्ती दो	८२
⋮⋮ Ụ	उमशांया अप्ती तीन	८३
⋮⋮ □	उमशांया अप्ती चार	८४
⋮⋮ ⌂	उमशांया अप्ती पाँच	८५
⋮⋮ ⌂	उमशांया अप्ती ६ः	८६
⋮⋮ ⬠	उमशांया अप्ती सात	८७
⋮⋮ ✶	उमशांया अप्ती आळ	८८
⋮⋮ ✶	उमशांया अप्ती नौ	८९
⋮⋮⋮	उमशांया नवती	९०

Surya Script Number System

आकृती	उच्चारण	अर्थ
⋯ –	अमशांया नवती एक	६१
⋯ L	अमशांया नवती दो	६२
⋯ ⌐	अमशांया नवती तीन	६३
⋯ ◻	अमशांया नवती चार	६४
⋯ ⌂	अमशांया नवती पाँच	६५
⋯ ⌂	अमशांया नवती ह:	६६
⋯ ✧	अमशांया नवती सात	६७
⋯ ✦	अमशांया नवती आठ	६८
⋯ ✹	अमशांया नवती जौ	६९
⋯	अमशांया सतम्	१००

आकृती	अर्थ	अंक
⋏	१,०००	$१०^२$
✚	१०,०००	$१०^४$
✳	१,००,०००	$१०^५$
✴	१०,००,०००	$१०^६$
✵	१,००,००,०००	$१०^६$
✺	१०,००,००,०००	$१०^८$
✺ (circled)	१,००,००,००,०००	$१०^९$
✺ (circled)	१,००,००,००,०००	$१०^{१०}$

114

Summary in Hindi

आकृती	अर्थ	अंक
✺	९,००,००,००,०,०००	१०⁹⁹
✺	९,०,००,००,००,०,९,००	१०⁹२
✺	९,००,००,००,०,००,०००	१०⁹३
✺	९,००,०,००,०,००,०,९००	१०⁹४
✺	९,००,०,०,०,०,०,००,००,०००	१०⁹५
✺	९,०,९,००,०,०,००,०,००,०००	१०⁹६
✺	९,०,००,०,००,०,००,००,००,०००	१०⁹६
✺	९,०,००,०,००,००,००,०,००,०००	१०⁹८

CHAPTER # 5

THE USES OF THE SURYA SCRIPT NUMBER SYSTEM

1) **Wall Clook**
2) **Calendar**
3) **Books Serial Number/Page Number**
4) **Calculation**
5) **Computer & Devices**
6) **General Number System**

The Wall Clock of the Surya Script Number System

Calendar of Surya Script Number System

Calculation of Surya Script Number System

Computer/Laptop Surya Script Number System

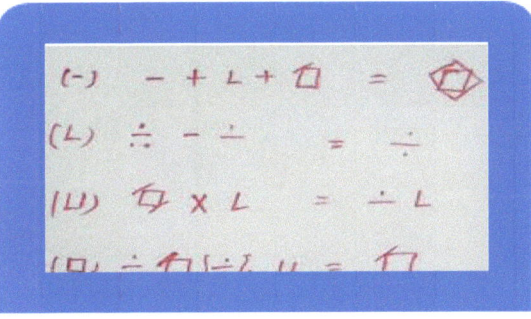

Books and Pages Surya Script Number System

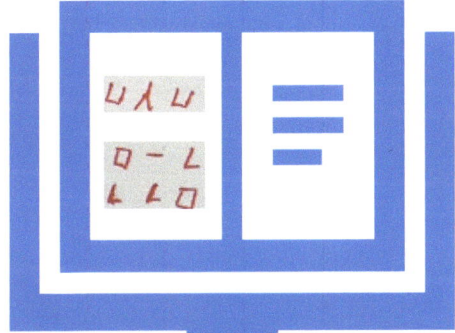

General Purpose Surya Script Number System

SECTION TWO

THE ONE-DAY SECRATE OF KAILASH

The One-Day Secrate of Kailash is dedicated to one of the unbelievable facts or truth but it's hidden for human beings.

Facts # 1

As we know a lot of people believe the Kilasha Mountn is not only a mountain but some invisible power is situated on the Kailasha. It's one of the points of research but we don't have any proof.

Facts # 2

Lots of researchers have tried to climb Kailasa but they never succeeded on his life. When they try to climb on Kailash his age increases suddenly and they feel very uncomfortable. As we know that lots of facts.

Story Point # 1

The Invisible power does not have any physical way like a human being. But they are more powerful than human beings. General Humans can't believed because their power is beyond the thinking of the human species. They can change very quickly beyond your thinking of the human mind.

Story Point # 2

As we know, some of the most powerful energy is situated in Kailasa. As we say, "Traiyanta." The Traiyanta has majorly three or four formations. The main formation's energy is situated on the south side of the palace. Inside, there is a palace with lots of very powerful energy residing in it. They have many cosmic weapons that are beyond human comprehension. When you descend one stair on the east side, the second and third energies are situated in Kailasa. Now, when you descend 10 stairs, the fourth energy is located in Kailasa. The fourth energy has its forces of medium and mixed energy. They have their forces. The fourth energy exists to protect the whole palace of Traiyanta.

Summary

As we know any type of energy and formation depends on your belief and faith. if we believe so we can do but if we don't so we always reject them. It all depends on your beliefs. if we believe this book is beneficial for us so we read if not then we leave that for our next destination.

ACKNOWLEDGMENT

I would like to express my deepest gratitude to everyone who has contributed to the creation and development of the Surya Script Number System. This project has been a labor of love, and it would not have been possible without the support and encouragement of many individuals.

First and foremost, I thank my family for their unwavering belief in me, for their patience, and their constant support throughout this journey. Your encouragement has been a pillar of strength.

A special thank you goes to my colleagues and mentors, whose feedback, critique, and expertise helped shape this book into its current form. Your thoughtful suggestions have significantly enhanced the depth and clarity of the material.

I would also like to extend my gratitude to the editorial team, designers, and everyone involved in the production of this book. Your dedication and professionalism have ensured that this work is presented in the best possible way.

Finally, to the readers—thank you for your interest in this book. I hope it brings new perspectives and a deeper understanding of the Surya Script and its profound impact on numeral systems.

With heartfelt appreciation,

Sujit Kumar Mishra

ABOUT AUTHOR

Sujit Kumar Mishra is an accomplished author, actor, and software engineer, whose work spans both fiction and non-fiction. With a unique blend of creativity and technical expertise, Sujit's writing draws readers into thought-provoking worlds, whether through captivating fictional narratives or insightful, real-world explorations.

Sujit has been recognized for his contributions to literature, winning the prestigious *Golden Books Awards* and *Prime India Awards*, accolades that celebrate his ability to connect deeply with audiences across genres. His writing aims to inspire, challenge, and entertain, offering something for every reader—whether seeking to lose themselves in a compelling story or to gain fresh perspectives on life and society.

When not writing or acting, Sujit enjoys exploring new technological innovations and immersing himself in diverse creative endeavors. Through his work, he hopes to foster a greater understanding of human emotions, societal change, and the limitless possibilities of storytelling.

For readers, Sujit's message is simple: Life is a story worth telling, and the journey of discovery never ends.

Also by (Plug Othres)

SL	Your Notes

www.ingramcontent.com/pod-product-compliance
Lightning Source LLC
LaVergne TN
LVHW071321080526
838199LV00079B/645